Business Writing Scenarios

Writing from the Inside

Business Writing Scenarios

Writing from the Inside

Jon Ramsey

bedford/st.martin's
Macmillan Learning
Boston | New York

For Bedford/St. Martin's

Vice President, Editorial, Macmillan Higher Education Humanities: Edwin Hill
Editorial Director, English and Music: Karen S. Henry
Senior Publisher for Composition and Business and Technical Writing:
  Leasa Burton
Executive Editor: Molly Parke
Developmental Editor: Beth Castrodale
Associate Editor: Rachel C. Childs
Editorial Assistant: Evelyn Denham
Production Editor: Lidia MacDonald-Carr
Assistant Production Supervisor: Victoria Anzalone
Associate Marketing Manager: Sophia La Torre-Zengierski
Project Management: Books By Design, Inc.
Director of Rights and Permissions: Hilary Newman
Senior Art Director: Anna Palchik
Text Design: Books By Design, Inc.
Cover Design: William Boardman
Composition: Books By Design, Inc.
Printing and Binding: RR Donnelley and Sons

Manufactured in the United States of America.

1 0 9 8 7 6
f   e   d   c   b   a

For information, write: Bedford/St. Martin's, 75 Arlington Street, Boston, MA 02116   (617-399-4000)

ISBN   978-1-4576-6707-7

Preface

Human beings organize themselves into complex social, cultural, political, and business units. Effective communication practices are essential for sustaining these social, civic, and professional structures and for advancing their interests. At its heart, *Business Writing Scenarios: Writing from the Inside* is about business communications in this broader sense. The book centers on the human relationships facilitated or negotiated through written (and other) communications, which can help businesses thrive over time or, in other instances, leave them in a weakened state. *Business Writing Scenarios* also examines the value systems, thinking patterns, and communication strategies that are the hallmarks of successful leadership across a variety of organizational structures.

Central to this book are the scenarios of its name: real-world business situations that students are likely to encounter as they pursue their careers. The scenarios immerse students in challenging writing situations—for example, having to convey disappointing news, to explain a major policy change, or to respond to a difficult customer—and show how writers addressed these situations effectively or ineffectively. After analyzing the scenarios, students apply what they've learned through activities ("applications") that ask them to respond in writing to similar business situations. The applications provide advice on how to address the particular challenges of each writing situation, helping students focus on their purpose and audience and on the most productive communication strategies. As they analyze the scenarios and complete the applications, students build the thinking and writing skills that will serve them for years to come in any professional situation.

Business Writing Scenarios also helps students avoid writing gaffes, which can have serious consequences and which, thanks to the rise of digital communications, have never been easier for others to discover and disseminate.

Throughout the book, students will learn to communicate clearly and without wasting words and, thus, their audience's time. They will also learn to develop nuances of tone, learn how to marshal sufficient evidence, and learn how to understand audience psychology—in other words, to assess what will achieve a particular purpose based on audience needs and expectations. All of these skills are essential in complex business writing situations. This approach to business writing, and the focus of this book, asks students to imagine their way into business and leadership roles that most of them have not yet experienced and to think creatively and strategically as they write their way toward persuasive communications.

Get the Most Out of Your Course with *Business Writing Scenarios*

Bedford/St. Martin's offers resources and format choices that help you and your students get even more out of your book and course. To learn more about or to order any of the following products, contact your Bedford/St. Martin's sales representative, e-mail sales support (**sales_support @bfwpub.com**), or visit the Web site at **macmillanhighered.com /businesswritingscenarios/catalog**.

Choose from Alternative Formats of *Business Writing Scenarios*

Bedford/St. Martin's offers a range of affordable formats, allowing students to choose the one that works best for them. For details, visit macmillanhighered.com/businesswritingscenarios/catalog.

- *Paperback*   To order the paperback edition, use ISBN 978-1-4576-6707-7.

- *Popular e-book formats*   For details, visit **macmillanhighered.com /ebooks**.

Select Value Packages

Add value to your text by packaging one of the following resources with *Business Writing Scenarios*. To learn more about package options for any of the following products, contact your Bedford/St. Martin's sales representative or visit **macmillanhighered.com/businesswritingscenarios /catalog**.

LaunchPad Solo for Professional Writing
macmillanhighered.com/professionalwriting1e

Digital resources to enhance your professional writing course.

LaunchPad Solo for Professional Writing offers online tutorials on today's most relevant digital writing topics, from content management to personal branding, and allows students to work on whatever they need help with the most. Students develop the professional writing and communication skills they need to succeed both in the classroom and in the workplace, and can explore today's technologies in clickable, assignable learning sequences organized by popular professional writing topics. *LaunchPad Solo for Professional Writing* features:

Digital Tips. Step-by-step instruction for using technology to support workplace writing includes guidance for synchronizing data, assessing software and hardware, creating templates, and organizing productive online meetings.

Sample documents. A wide range of effective professional writing models provide students with e-mails, résumés, cover letters, reports, proposals, brochures, questionnaires (and more) to emulate.

Tutorials. Screen captures walk students through maximizing free online tools to access projects across platforms, design dynamic presentations, develop podcasts, manage their personal brand, and build common citations in APA and MLA styles.

Adaptive quizzing for targeted learning, skills practice, and grammar help. LearningCurve, a game-like adaptive quizzing program, helps students focus on the writing and grammar skills in which they need the most help.

The ability to monitor student progress. Use our gradebook to see which students are on track and which need additional help with specific topics.

LaunchPad Solo for Professional Writing can be **packaged at a significant discount**. Order ISBN 978-1-319-07313-8 to ensure your students can take full advantage.

Visit **macmillanhighered.com/professionalwriting/catalog** for more information. For technical support, visit **macmillanhighered.com /getsupport**.

Team Writing **by Joanna Wolfe**, University of Louisville, is a print supplement with online videos that provides guidelines and examples of collaborating to manage written projects by documenting tasks, deadlines, and team goals. Two- to five-minute videos corresponding with the chapters in *Team Writing* give students the opportunity to analyze team interactions and learn about communication styles. Practical troubleshooting tips show students how best to handle various types of conflicts within peer groups.

Join Our Community!

The Macmillan English Community is now Bedford/St. Martin's home for professional resources, featuring *Bedford Bits*, our popular blog site offering new ideas for the composition classroom and composition teachers. Connect and converse with a growing team of Bedford authors and top scholars who blog on Bits: Andrea Lunsford, Nancy Sommers, Steve Bernhardt, Traci Gardner, Barclay Barrios, Jack Solomon, Susan Bernstein, Elizabeth Wardle, Doug Downs, Liz Losh, Jonathan Alexander, and Donna Winchell.

In addition, you'll find an expanding collection of additional resources that support your teaching.

- Sign up for webinars
- Download resources from our professional resource series that support your teaching
- Start a discussion
- Ask a question
- Follow your favorite members
- Review projects in the pipeline

Visit **community.macmillan.com** to join the conversation with your fellow teachers.

Acknowledgments

While I have benefited greatly from the insights of many people who have published books and articles on business writing, *Business Writing Scenarios: Writing from the Inside* is based largely on my longtime experience as an administrator who faced countless organizational opportunities and challenges that required a written explanation, proposal, or defense. At least as important as the experiences I brought to this book are the wonderful administrative and faculty colleagues with whom I have worked over the years. Their wisdom, equanimity, clear-minded assessments of complex challenges, commitments to improving the organization and the lives of the people who worked in it, and superb writing abilities have enhanced my communication skills tenfold. The college and university students who have enriched my teaching experience have also made huge contributions to my understanding of what works in the classroom and what does not. They have graciously given me permission to include some of their business writing in this book, and their contributions are acknowledged by their names.

In particular, I want to thank my former colleagues at Skidmore College, especially Phyllis Roth, Ann Henderson, David Porter, Fran Hoffmann,

Susan Kress, Terry Diggory, and Sarah Goodwin. In their leadership and teaching roles they have been an inspiration. They probably never knew how much I learned from them about thinking, writing, and leadership. In their various roles they conducted a good deal of "business" for the college, and they always did so with deep respect for the views of others, with an unflagging commitment to fostering the interests of the entire organization, and with a quality of insight that time and again led us in the best direction during both prosperous and challenging times. Skidmore College also generously provided a grant to support my research for this book.

My more recent teaching, in the Writing Program at the University of California, Santa Barbara, has especially enriched my understanding of business writing pedagogy. My generous-spirited and talented colleagues in the Writing Program have given me every imaginable encouragement for my teaching and various academic projects. The fabulous mentors who enhanced the effectiveness of my business writing courses have included, especially, Jeff Hanson, Gina Genova, Janet Mizrahi, and Patrick McHugh, and on countless occasions Chris Dean has shared his teaching insights with me. Indeed, all of my colleagues in writing at UCSB deserve my deep gratitude for sustaining a truly supportive and stimulating work environment and for caring so deeply about our students' education. The inspired leadership of Linda Adler-Kassner has held all of these great qualities together for all of us. I also appreciate the help of my colleague Auli Ek and her students, who supplied the sample business plan for Chapter 7.

I also want to acknowledge my appreciation of the reviewers who commented on my proposal and manuscript: Bethany Lee, Purdue University North Central; Huatong Sun, University of Washington–Tacoma; Jennifer Veltsos, Minnesota State University, Mankato; Rochelle Brooks, Viterbo University; Stephan Byars, University of Southern California Marshall School of Business; and one anonymous reviewer.

Thanks also to my students, who contributed many of the examples in this book: Brittany Berin, Cameron Brinkman, Alexandra Kambur, Andreas Nitsche, Courtney Steele, Michael Cipriano, Sheena N. Joseph, Alyssa Kianidehkian, Andrea Michaelian, Gregory D. Leyrer, Lisa H. Newton, Nick Kohan, David Love, Jennie Stodder, Scott Pantoskey, Ally Diamond, Laura J. Francis, Martha Grimes, Xiangdi (Sandy) Wang, Daniel Levens-Lowery, Jenna Thompson, Kara Gorman, Anne Holston, Maher Zaidi, Nicolas Tomei, and Hovig Axle Wartanian.

The guidance the staff at Bedford/St. Martin's has provided over the last couple of years has been extraordinary. Thanks go to Leasa Burton for seeing value in this project in its earliest stages. I have been blessed with very discerning editorial advice from Beth Castrodale and Kate Mayhew, who on hundreds of occasions have improved my writing, filled in my blanks, and made expert strategic recommendations on the scenarios and other educational apparatus of the book. Evelyn

Denham and Rachel Childs shepherded the book with great skill through its final phase. Thanks also to Lidia MacDonald-Carr and Nancy Benjamin for guiding the manuscript through production, to Kathleen Benn McQueen for the meticulous copyediting, and to Janis Owens for the design work.

Finally, I want to thank my wife, Kitty, not only for her constant encouragement as this project moved forward but also for her very insightful corrections and clarifications. Throughout my professional life she has been one of my most trusted guides.

About the Author

Jon R. Ramsey was an Associate Professor of English and the Dean of Studies at Skidmore College until 2004. His career in administration and teaching continued through 2014 at the University of California, Santa Barbara, where he was the Director of Policy and Publications for the Graduate Division and a continuing lecturer in the UCSB Writing Program. He has published a number of articles and book chapters and has co-edited two books on literature, writing, and administrative issues. As an administrator and office director he has been especially involved in the creation and implementation of new programs in the United States and abroad and the construction of a wide variety of policies frequently requiring complex written negotiations with a myriad of internal and external audiences. He earned his BA at San Diego State University and a PhD at the University of California, Riverside.

Contents

2   An Overview of Business Writing 20

Before you begin composing any piece of business writing, you will need to consider your role and authority within the business organization, your purpose in writing, the audience you must reach, the strategy that is most likely to achieve the desired goal, and the best medium for your communication purposes. This chapter examines these important factors.

Knowing how to draft an effective résumé and cover letter is key to your success in securing a position and advancing in your field. This chapter gives practical advice on creating these important documents—and on succeeding in the job search.

4 Business Document Design, Formats, and Conventions 83

As a business writer you want to create documents that are as aesthetically pleasing as they are easy to read and understand. This chapter covers design basics, as well as important document formats and conventions.

5 Writing to Colleagues within the Organization 104

It is important to address colleagues as valued partners in a shared enterprise. This style of communication will help foster a culture of cooperation, productivity, and respect that is crucial to any organization's success.

Understanding the Challenges of Writing to Colleagues 104

Responding to Real-World Writing Scenarios 112

6 Writing to External Constituencies 127

External audiences with whom you communicate will view you as a representative of your organization, so it is important to make a good impression on them. This chapter will help you communicate effectively with such audiences, even under the most challenging circumstances.

7   More Complex Business Writing Projects   159

As you advance in your career, you may be called on to develop longer, more complicated documents, such as grant proposals or business plans. This chapter will introduce you to the fundamentals of these more complex pieces of writing.

8 Business Writing Gaffes in the Real World 218

Even highly experienced business professionals make communication errors at times, and you can learn from their mistakes. This chapter provides examples of some especially glaring gaffes to avoid.

9   Leadership Values in Business Writing   241

In this chapter, we'll take a closer look at leadership qualities that you can foster in yourself to become both a better communicator and a better colleague.

CHAPTER 1

The Purposes and Concepts of This Book

Understanding the Nature of Business Writing

You already engage in business writing in many of your routine and more formal interactions with others—you just might not know it. You might, for example, write to a faculty member about crashing her course or about an upcoming assignment, negotiate by e-mail with a team member in your class, or lobby in writing against anticipated tuition hikes at your university. In all of these situations you are considering what you want to make happen, the audience you are addressing, and the best strategy for writing clearly and persuasively.

Even to *enter* into the world of work, you need to rely on your writing skills to craft effective cover letters and résumés and to communicate effectively with representatives in human resources. Once you've been hired and you become a member of a professional community, you then might have to communicate persuasively with an unhappy customer or client, provide clear financial information to a supervisor, or help write a proposal for a new product or service. Whether in person, in print, online, or over the phone, it's important to consider the tone and message you wish to convey, your audience, and the appropriate medium for communicating that message.

Previewing the Basics

To begin our discussion of what distinguishes effective business communication from poor business communication, let's start with some everyday examples.

An Example of Unspecific, Unfeeling Business Writing

During the summer of 2012 I stayed for three days in a London apartment. On my first morning in the apartment there was no hot water, and I had to endure a cold shower. I wondered, was this a problem for the individual apartment or for the entire apartment building? I made my way to the lobby, where the supposedly 24/7 porter was not on duty. Several long-term tenants had also gathered there to learn the cause of the problem. Posted to one side of the front door was a handwritten message:

The plumbing problem will be fixed tomorrow.

There was no date or time on this concise message (which none of us had seen posted until that morning), no way to determine whether "tomorrow" meant today or the next day. And even if the next day were the target repair date, if the plumber arrived in the afternoon we would all face yet another morning of cold showers. The other residents in the lobby expressed various degrees of amusement and anger about the situation. Given the ambiguity, I was prepared to abandon my apartment and move into a B&B hotel around the corner.

An Example of Specific, Empathetic Business Writing

Had the building manager added some context to the plumbing notice and shown some empathy for our situation, the other tenants and I would have felt reassured that the problem was being addressed:

Plumbing Problem — Wednesday, September 21
We are very sorry for the lack of hot water this morning and have notified our plumber. He believes he can repair the boiler and restore the water supply before noon today. Any further updates will be posted in the lobby.

The revised plumbing notice would have prevented a lot of confusion and irritation. Both examples should remind you that the crafting of clear communication is not just reserved for major documents. The author of the terse note did not think about the likely frustrations of his or her audience, did not provide a context for the problem or a time line for the repair, and never hinted at any sympathy for the shivering tenants; thus, the attempt to communicate was comical at best. The manager's poor communication could hurt business; because the tenants had no idea when the plumbing would be fixed, some of them might have decided to leave and stay elsewhere. (By the way, the end of the London story is this: The plumber rang the phone in the unattended lobby, which I decided to answer, and he assured us that he would attempt to repair the problem that very day, but he needed the porter to unlock the basement. The porter did show up a couple of hours later, and the plumbing was repaired.)

The Foundations of Effective Business Writing

Whether you are a student learning more about business writing, a professional in the field, or a business owner who needs to let his or her tenants know that the plumbing is on the fritz, it's important that you step *inside* the situation in order to craft an effective response. The scenarios in this book will help you do just that: by entering these real-world business situations, you will learn to carefully examine the circumstances you are facing, identify the purpose or goal of your communication, consider your audience and the best medium to use when communicating with them, and anticipate the long-term impact of fostering good relationships within the workplace as well as with clients, customers, and other external constituencies.

If you are like most undergraduate students, you have held jobs in retail, at restaurants, or in deliveries. Some of you may have held office positions of more complexity and responsibility and can use your experience to good advantage in a business writing course. Yet even those of you with office experience might not have written business communications as a company representative—documents that can have an enormous impact on an organization's reputation or prosperity.

Regardless of your employment history, you have had to communicate with friends, family, and teachers on complicated matters. You have probably also interacted with happy or disgruntled people in a job setting, whether online, over the phone, or in person. In all of these social or business interactions you needed to take into account your purpose or goal in speaking with or writing to another person, your tone of voice, the persuasive strategies you could use, and the expectations of your audience. This common set of experiences actually provides a good foundation for many of the challenges of business writing.

In daily communications, most of what we say (or perhaps tweet or text) is pretty spontaneous, not planned out carefully. That is one big difference you must consider for effective business writing: professional writers need to think carefully about **their purpose in writing** (what they want to achieve), **the role they occupy in their company or organization** (for example, their level of authority and areas of responsibility), **the situation, or scenario, confronting them** that requires a piece of writing, and the often elusive **psychology and strategy that will work** well for a particular audience. As they become more experienced, business writers comprehend these complexities more quickly and with more assurance. That should be your goal as well: to increase your familiarity with various business genres and the writing strategies that are likely to get the desired results in different situations.

Imagining Your Way inside Business Situations

If you have limited experience with writing for business purposes, you will need to imagine your way into another world—to write from the inside. You will need to play "what if" and "let's suppose" in order to respond effectively to the business situations you are likely to encounter later in your professional life. Even experienced business professionals perform these imaginative acts every day, but over time the process becomes automatic for them—as it will become for you. And as mentioned earlier, you can also draw on many of your personal experiences and interactions when considering how to respond in many of these business situations.

The Scenarios in This Book

The business situations, or scenarios, presented throughout this book will ask you to step outside the limiting boundaries of your own experience and put yourself in the shoes of an insider who is facing an actual business task. In the "application" activities accompanying the scenarios, you will be asked to assume, for example, the voice and outlook of a customer-relations representative, a human-resources assistant, an assistant director of finance, or even a chief executive officer (CEO). While most of the scenarios in this text are hypothetical, very soon you will be immersed in a career in which these or similar situations might become commonplace.

In responding to the scenarios, you will write as part of, or as the person in charge of, a decision-making process, in every case considering the audience and the best possible method for communicating (such as through e-mail or by memo) before crafting a response. You will be challenged to negotiate, explain, defend, request, appease, apologize, or promote a cause through your professional writing.

Because they are intended to build strong business writing skills, the applications in this book offer no easy way out. For example, you cannot pass a business problem onto someone else in the company, you cannot claim ignorance to make an unhappy client go away, and you cannot postpone bad news that will just make the recipient of the news even more displeased later on. In every case, you will have to take responsibility, even if you or your company is at fault and even if a cloudy or euphemistic response might make you more comfortable for the time being.

As you study the scenarios and respond to the applications, keep in mind that in an actual business setting the content and character of your writing can have a significant effect on how colleagues, customers, and clients perceive you and the company as a whole. Through clear, effective writing you may garner respect and opportunities for growth within a company and beyond. Serious errors, however, could lead to a demotion or even to termination.

In addition, in your business writing you will have to consider your level of authority and role in the organization, company policies, ethical and legal issues, and the possible effects on both your reputation and your company's, positive or negative, if you write with a particular tone and strategy. You will always have to anticipate the needs and concerns of your audience.

Let's prepare to step outside your more familiar personal and business experience by looking at some effective responses to business writing scenarios.

Sample Scenario: A New Employee Introduces Herself

In this scenario, a student worker has just been hired to provide computer support to a 39-member university writing program, and she wants to introduce herself and inform staff of her schedule. Here is the e-mail she sent to the faculty as they prepared to return for the fall 2016 semester:

Hi everyone,

My name is Brittany Berin, and I'll be your Computer Tech Support for the upcoming year. I just figured out my work schedule and wanted to let you all know that I'll be stationed in the 1515 computer lab every week this quarter at the following times:

Monday	10:00 a.m. – 12:00 p.m. & 1:00 p.m. – 3:00 p.m.
Tuesday	2:00 p.m. – 3:00 p.m.
Wednesday	10:00 a.m. – 12:00 p.m. & 1:00 p.m. – 3:00 p.m.
Thursday	2:00 p.m. – 5:00 p.m.

---

continued

Please feel free to stop by the lab at those times if you have any questions or computer troubles for me, and, of course, I'm always available by e-mail.

I'm looking forward to working with you this year.

Brittany Berin

The author of this message—the new tech-support specialist for the coming year—knew that the faculty she was addressing were pretty informal and would treat her as a colleague, so she could afford to be casual in her communication. Note how clearly she presented the information needed for the faculty. She also chose to introduce herself through e-mail—a good tactic for a tech-support specialist because now those instructors who might need her assistance throughout the year have her e-mail address. This simple e-mail presents the writer as a person who is organized, clear-minded, and eager to fulfill her job responsibilities. By carefully considering her audience and the role she will play in the larger organization, she has already made a positive first impression.

Sample Scenario: A Manager Politely Declines a Colleague's Request

Even casual communication between colleagues can present some challenges. Consider this fictional scenario: JT is writing to you, a manager with whom he has a friendly relationship in the office (though only in the office setting). You have received a confidential draft memo from Finance because you are a manager at the company (JT is not). The memo outlines emerging plans to "reconfigure" the workforce at Armitage Brothers, which means that some employee positions may be redefined in status and authority, assigned to other areas, or even lost. JT wants you to share this information so that he might plan ahead for the worst-case scenario:

Hey there,

Could you get your hands on the memo that old Birtie in Finance circulated to the managers last week? I know it's supposedly hush hush, but I need to see whether the position of my assistant is going to be eliminated as we move toward the stupid "reconfiguration" of the labor force at Armitage Brothers. If you don't have the memo, maybe you know who does? Thanks a million.

JT

Consider how you might respond in this situation: To start, who is JT as an *audience*? Unlike you, JT is not a manager and was not involved in reviewing the confidential issues under consideration. On the other hand, JT is a valued colleague, and you want to retain a good working relationship with him. For this reason, even as you decline his inappropriate request, the *tone* of your response should be respectful. Your *role* is that of a manager, and you need to address JT as a valued colleague while also honoring your relationships with other managers (in this case, an agreement of confidentiality).

It might seem tempting simply to give JT the information he has requested, but consider some of the possible consequences. You can be pretty sure that other employees and managers will find out that you leaked confidential information. Some staff will then seek the same favor from you, and your fellow managers will stop trusting you to participate in decision-making processes. Further, the early release of the draft document might send shock waves, unnecessarily, through the staff ranks.

Then again, you might just tell JT that you have not seen the memo. But that white lie will be easily discovered when JT approaches another manager. JT will no longer trust you and might not work as cooperatively toward your shared goals in the future.

Another approach would be to just kick the can down the road, so to speak: send JT back to "old Birtie" or to another manager for the information. If you use this strategy, however, your manager colleagues will resent that you took the easy way out, pushing the issue onto them. You will have failed in your leadership, in your managerial role.

Still another tactic would be to adopt JT's jaunty tone in a friendly reply, echoing JT's assessment of the "stupid" planning process (to which you contributed) and blaming the confidentiality on "old Birtie." If you fall into this trap, however, you can bet that JT will share this juicy e-mail with someone else, and pretty soon Birtie will discover your true colors, which will only damage your reputation as a manager and a co-worker.

As you can see then, there are a number of potential pitfalls in this situation. You must embrace your role as a manager and decline JT's request while somehow retaining his allegiance and respect. The approach you take should imply your friendly concern and also provide the basic reasons why the report needs to remain confidential until the decision-making process is completed. At the same time, you should not preach ethics to JT, which could provoke his resentment.

On the following page is one example of a successful student response.

Hi JT,

I appreciate the fact that we have been good pals here at Armitage Brothers for quite some time now, and I would love nothing more than to help out a friend in need, but unfortunately my hands are currently tied. I've been given strict instruction from Finance to keep that memo confidential until all the details of the reconfiguration have been concluded. I realize this process can be somewhat painful, and I understand your desire to plan ahead, as the work environment here at Armitage Brothers has grown a bit hectic recently due to the various reconfiguration rumors. That being said, I can reassure you that management is currently doing all it can to conclude the details of this reconfiguration in a timely fashion. Again, I wish I could divulge more at this time, but my responsibilities require my silence. I will be sure to fill you in on the details of this matter just as soon as I'm permitted.

Best Regards,

Cameron Brinkman

Notice the writer's friendly tone balanced with firm resolve, his sympathy for JT's anxiousness, his respectful characterization of the reconfiguration process, and his nonpreachy explanation of his reasons for saying "no." The strategy used is essentially "I wish I could, but I can't because . . ." This e-mail meets Cameron Brinkman's managerial obligations and will probably retain JT's respect, but it sends a firm and friendly signal to JT regarding future company processes.

Writing to Build and Maintain Relationships

The previous scenarios highlight a central tenet of effective business writing: the importance of interpersonal considerations. Whether the purpose is informational or persuasive, *business writing always involves people writing to other people*. While a computer algorithm might be responsible for generating millions of "thank you" or "sorry" messages to, for example, Amazon or Verizon customers, it was a person or a group of people who originally crafted the message. When you receive these generic notices, you know you are not being addressed personally, but you do care about the tone and clarity of the messages. And well-run companies care a great deal as well. (See Chapter 8, "Business Writing Gaffes in the Real World," for examples of how smart professionals sometimes fail in their mass or individual communications.)

In every business writing situation, you will need to ask yourself how you will achieve your purpose without alienating your audience. How, for example, do you take a firm stand with a co-worker without disrespecting or insulting him or her? How do you convey bad news to colleagues in a way that says "we can face this challenge together"? How do you persuade members of your organization to move in a new business direction without slighting established company traditions or practices? How do you respond to an angry client or customer with honesty and reassurance but without exposing the organization to further liabilities? How do you tell the story of your organization effectively in a grant-writing process? How do you request a special service for the company (for example, a keynote address or an endorsement) from a well-known, influential person? These are but a few of the daily tasks of writers representing business organizations.

In the case of JT's request, you were asked to resist the temptations to pass the buck, to fabricate face-saving (and only temporarily effective) excuses, or to respond in a euphemistic or vague way. It would have been easy to simply say: "Sorry, but I can't give you that information." In a busy professional setting, sometimes such abrupt or dismissive communications are necessary. Many professionals have found, however, that their addressing colleagues and customers with concern and respect builds relationships that have an enormous positive impact on how individual professionals, and the company's overall character, are perceived.

Yet there will be times when you do need to hang up the phone, compose a terse and unequivocal message, or reply to some number of people with the generic "we appreciate your concerns and will give them all the attention they deserve." For example, a customer might have repeatedly and aggressively made his point, and you need to end the ongoing interactions. Or you might need to make clear that a company decision is final and needs to be implemented immediately, regardless of whether the employee or colleague approves.

So "whenever possible" is a good rule to follow for the respectful tone recommended throughout this book for business communications and for human relationships in general. Whether addressed to colleagues within your organization or to customers or clients outside it, respectful writing (and speaking and behaving) is good business. Even when the situation is tense and contentious, you should try not to burn bridges—*whenever possible*.

Choosing the Medium

In addition to identifying a communication strategy that suits your purpose and audience, you need to choose the communication medium most appropriate for a particular business situation. Specifically, you should consider when a person or group is best addressed by a telephone call, a printed letter or memo, an e-mail (with or without an attached document), a text message, or some combination of these methods.

Factors to Consider

Sometimes the urgency of a situation, or the importance of the colleague or client with whom you need to interact, suggests that a phone call, which is more prompt and personal than other methods of communication, is indicated. You might then follow up with an e-mail or a pertinent document because the information you need to convey is too detailed or complicated to provide over the phone. In other situations, a more formal printed document will serve your purpose better—perhaps to establish a genuine paper trail for contractual or legal reasons, or just to suit the convenience of the receiving party.

These questions of the best medium to use might also depend on the personal preferences of the recipient. You might have a supervisor, for example, who prefers to receive digital versions of nearly every communication or document. Or the company for which you work might have established policies on the formats or media to be used for different genres of communication.

Any professional whose work has spanned the past 20 years or more would testify that e-mail and digitally transmitted documents have rapidly displaced at least 80 percent of the hard-copy communications that used to fill endless binders, file cabinets, and company warehouses. (It's interesting to consider that a $10 flash drive would now probably hold all of the text ever produced by a businessperson over his or her 30- or 40-year career.)

Examples of Selection Strategies

If the organization for which you work does not have guidelines on when to use a particular medium of communication, the choices will depend largely on your growing experience with various audiences and scenarios. Here are examples of how you might choose a medium based on various purposes, audiences, and situations:

- A printed invitation might be mailed to a keynote speaker for your company's upcoming celebration and then followed up with a phone conversation to explore expectations with the invitee.

- An important client who is unhappy with his recent order might require an immediate phone call followed by an e-mail with a further apology and reassurance.

- An announcement of revised sick-leave or travel-reimbursement policies might be introduced in a brief e-mail to all employees but with the actual policy document circulated as an attachment to the e-mail.

Writing to Achieve Your Purpose—and Get Results

Much of what you have written in college has been produced for individual courses or instructors and for academic purposes appropriate to various disciplines. Consequently, the impact of your writing has generally been limited to teacher feedback and grades. In a business writing course, you are still writing to meet your teacher's expectations, and it is likely that your audience will also include your peers. But your most crucial audience will be the colleagues or external constituencies specified in writing assignments, assignments like the "applications" included in each chapter of this book. These activities will give you practice in writing to the types of readers that you will encounter in the real world of work—for example, a potential investor, a favor-seeking client, an angry customer, an invited dignitary, or an anxious colleague. While your teacher remains an important audience, he or she will be assessing your effectiveness in addressing business-related issues and audiences.

In each application, you will gain practice anticipating the consequences for your choices as writers that extend beyond the university's more traditional academic outcomes. You must carefully consider your audience, the most effective tone to adopt, the organizational goals to be achieved, and the most essential information and evidence to deploy to achieve the goals of all involved.

Some of the purposes or goals that inform different genres of business writing are sketched below. While the list might appear long and daunting at first glance, you'll find that many of the categories overlap, and some of the examples could fall into more than one. This list of purposes will suggest what you will need to determine before you compose a business communication—or at least before you press "send."

Explaining, clarifying, or providing information. The writer's purpose here is not to persuade the client or customer but simply to provide clear information. For example, the writer might want to

- explain to potential investors just how mutual funds differ from exchange-traded funds (ETFs).

- clarify the policies and procedures for employees requesting travel reimbursements.

- explain the different types of medical plans available to employees.

- provide clear directions for assembling or activating the company's product.

- compose a fact-based research report on employee productivity over a period of time.

Defending. Here, the writer's purpose is to defend a company policy, practice, or decision. For example, the goal might be to

- defend the organization's practices or policies against an emerging legal challenge.
- defend a colleague who has been wrongly faulted.
- justify a controversial financial decision to a board of directors.

Negotiating. This is an element of many types of written communications, but especially those seeking to settle an area of dispute or to resolve competing claims and interests. For example, the writer might want to

- grant or decline employee requests for promotion and to do so according to consistent criteria.
- seek an agreement between employees and administration regarding a disputed policy on sick leave.
- state clear guidelines that delineate exactly when, by whom, and by what means a contract can be renegotiated.

Appeasing, mollifying, ameliorating, or apologizing. This is a frequent necessity of many organizations in their efforts to sustain a company policy or practice in the face of special concerns from important constituencies. For example, the goal might be to

- decline a "legacy" college admissions request from an influential alumna or alumnus.
- retain the loyalty of an important customer when he or she has a complaint concerning company service.
- retain the commitment of a valued employee while declining her or his promotion or vacation request.
- offer a special favor to an aggrieved customer as compensation for an actual or perceived company error.

Persuading, requesting, or marketing. Here, the writer's effort is to present organizational goals in a positive light to achieve a particular outcome. The purpose might be to

- gain the interest of a prospective employer.
- invite a distinguished person to a company event.
- acquire funds through grant writing.
- construct an accurate and inspiring mission statement.
- market the products or services of a company.

Constructing policies and procedures. This is a type of explaining and persuading that defines organizational goals (and potential missteps) and devises a pathway for achieving the desired ends. Policies and procedures tend to be used within an organization, and they might seek to

- describe employee eligibility for company medical and other benefits.

- clarify travel-reimbursement policies and procedures.

- construct grievance procedures.

- devise guidelines for the personal use of company phones, office machines, and Internet connections.

Proposing new or modified organizational practices and directions. In the same vein as constructing policies and procedures, the purpose of proposing modifications is to shift important company decisions in a particular direction and to promote new initiatives within the organization. These communications might seek to

- propose a new program that benefits the health of employees while also reducing the cost of medical benefits.

- gain traction for improving the company's product or service.

- make a case for reorganizing the workforce in a way that allows for more flexible time commitments.

- move colleagues toward a new business model that responds to customers' environmental interests.

Responding to Real-World Writing Scenarios

Let's apply a few of the purposes and strategies outlined previously to the following scenarios.

Analyzing a Writing Scenario: Explaining a Workplace Disruption

Imagine that the water in your workplace needs to be turned off for two full workdays, and the CEO (you) needs to alert employees of this disruption, anticipate what impact the announced water shutoff will have, and plan ahead for employee needs and concerns.

The Challenges of Explaining the Disruption

Since this situation affects the entire workforce, not just one or two individuals, you will need first to do some strategic thinking: what are the definite and possible effects of there being no water available, and how exactly will you accommodate your colleagues' needs and comfort during the days without water? After you have figured out these impacts and the needed remedies, you can craft a memo that both acknowledges the inconveniences that will be created and reassures employees that you have anticipated their needs and concerns.

An Ineffective Explanation of the Disruption

The following CEO memo condescends to employees, placing the author and other "bosses" on a level above them; it fails to anticipate the problems that will arise for employees working without drinking water or lavatories; and it only vaguely describes the need for this interruptive repair:

Digital Services Incorporated
Fox Creek, Idaho

DATE: September 19, 2016
TO: All Employees
FROM: [Your name], Chief Executive Officer
Subject: Water Shutoff

The bosses and I need to alert you that the water will be turned off in our section of the building this coming Wednesday and Thursday to remedy a major plumbing problem. We are sorry for the inconvenience and know that you will meet your work responsibilities as usual. Any special problems concerning the water shutoff should be reported by tomorrow, Tuesday.

Have a nice day.

Applying What You've Learned

The following activity asks you to improve the Digital Services memo above by applying the advice offered earlier in this chapter.

APPLICATION 1-A

Explain a Workplace Disruption

Write a much friendlier and more strategically sound memo than the one just presented. Here are some tips:

- Assuming you are the CEO, first think through the actual effects of having no water available for two full workdays and devise strategies for addressing these problems before they arise.

- Consider how to develop a tone that treats employees as colleagues, not as peons being ordered about by the bosses. You will need to compose a communication that relays sympathy for the workers as they face the water shutoff.

- Clarify any vagueness surrounding the reasons for the water shutoff. Employees will be more understanding of the inconvenience if they know that there is a serious need to repair the plumbing during work hours and that, in the long term, the repairs will benefit them.

Here are further suggestions on how to craft a memo with better tone and improved strategic planning:

- Plan ahead for the various consequences of the employees' having no water in the usual drinking fountain, lavatory, and break room. Your simply asking them to do without the drinking water and lavatories would not meet ordinary health and labor codes and would, obviously, make for an impossible two workdays.

- Consider the level of detail you need to describe the plumbing problems in order to convince employees that the shutoff is necessary (some might suspect management bad faith or incompetence). Adults want to know *why* things are necessary; they are not satisfied with "just because" assertions.

- Make arrangements with a business in the same building to use its lavatories and drinking fountains. Supply bottled water or bring in portable water coolers. If possible, provide exact locations for these resources so that you don't need to send out a second memo on "where" and "what."

- Ask if any employees have special needs regarding drinking water (for example, taking medications or needing frequent hydrating) or require frequent lavatory visits. Provide the name and phone number or e-mail address of the person employees should contact with any special needs or concerns. (Don't add to colleagues' irritation by making them look up the phone number or e-mail address of "Ed" in Human Resources.)

In addition to good strategic planning on the part of the CEO, considerations of tone and audience are also very important for this memo to be well received. Certainly get rid of the off-putting reference to "bosses" used in the example of an ineffective memo; instead, address the staff as valued colleagues. Consider providing an employee "perk" of some sort on the two affected days in order to boost morale, or perhaps allow certain staff to work from home. Finally, note the cheery, illogical closing of the CEO's memo. It's not a good idea to mix tonal modes, in this case presenting bad news and then ending with a seeming non sequitur: the water-shutoff days will not be a terrible problem if handled properly, but they will not be especially "nice" days. A cheerful closing will just irritate your audience.

An Effective Explanation of the Disruption

Here is an example of how a student of business writing recast the water-shutoff memo effectively. The writer presents herself in a relatively friendly way and has already figured out how to provide the necessary resources for the two days without water:

Digital Services Incorporated
Fox Creek, Idaho

DATE: September 19, 2016
TO: All Employees
FROM: Alexandra Kambur, Chief Executive Officer
Subject: Water Shutoff: Wednesday (Sept. 21) and Thursday (Sept. 22)

Dear Colleagues:

Last night pipes broke in both of our restrooms, causing significant water damage that requires immediate attention. Our maintenance crew informed us that they will need to turn off the water in our section of the building this coming **Wednesday and Thursday (Sept. 21–22)** in order to fix the major plumbing problem.

Unfortunately, on Wednesday and Thursday the restrooms and water fountains in our section of the building will be closed for repairs.

Alternative Copy, located on the first floor of the building, has generously offered to let us use their restroom facilities while ours are being repaired. To find the restroom, take the elevator down to the first floor and turn right. The men's and women's restrooms are located down the hallway before the large entryway into Alternative Copy. Drinking fountains are right next to the restrooms.

We recognize that this is an inconvenience and will be providing beverages and snacks. There will be bottled water, sodas, coffee, tea, bagels, and fruit in our break room all day. In addition, we will be placing a bottle of hand sanitizer in the break room and at the front desk for your use.

If you have any questions or concerns, please contact Joe Pieper at extension 2255 or at jpieper@gmail.com. Joe will be happy to answer any questions and arrange any necessary accommodations.

We are very sorry for the inconvenience and greatly appreciate your patience and cooperation.

Sincerely,

Alexandra Kambur

The student's version of the memo is friendlier than the original, defines the plumbing problem more clearly, and shows the CEO planning ahead for the needs and lingering questions of her colleagues. She makes unmistakably clear when the water will be turned off (for example, there is no ambiguous reference to "next" Wednesday and Thursday).

Notice that she has used the plumbing problem itself as the "buffer" to the central negative news, the shutdown of the water supply for two days. She has arranged for alternative water and restroom facilities, provided specific directions to these facilities, and named a staffer whom employees can contact if they have any further concerns or special needs for the affected workdays. She has also decided to offer snacks, sodas, bottled water, coffee, and tea to support the morale of her colleagues. Overall, the memo demonstrates not only the student's ability to communicate clearly and with care and respect for her colleagues but also her ability to anticipate the needs of her colleagues and to arrange for remedies. She has gotten completely inside the business situation.

Analyzing a Writing Scenario: Responding to a Former Colleague Who Wants Confidential Information

Kenneth Sprocket is an old friend through a company you used to work for, Feldstar Electronics. Feldstar is rumored to be merging with your current company, Technic Ltd., and Sprocket wants a heads-up on the details of the merger negotiations. Can you help him? What could be the consequences if you gave him the confidential information he is requesting? If you decline to share this information, how can you do so without offending an old friend and colleague? What else is possibly at stake here?

Dear Jill:

You know I have not asked many favors of you, and our business relationship goes way back. I hope you can help me out with this. I need some tips on what your firm is going to want to know at our joint meeting tomorrow—just so we can get our ducks in a row. I realize this is only a preliminary meeting on our possible merger, but I think it would aid the process if we shared some financial and other information beforehand. Whatever you can provide will be greatly appreciated. Thanks a million.

Your colleague-to-be,

Kenneth Sprocket

Applying What You've Learned

Considering the advice and examples presented earlier in this chapter, complete the following activity.

APPLICATION 1-B

Respond to a Former Colleague Who Wants Confidential Information

Write a response to Kenneth Sprocket, considering the serious—perhaps even legal—consequences if you were to share confidential company information in advance of the official merger discussions. How can you decline Sprocket's request without offending him, particularly given that you might soon be working for the same (merged) company? For insights on how to respond tactfully, you might refer back to Cameron Brinkman's communication with JT (see page 8).

Looking Ahead

The following chapters will present more scenarios and activities that will immerse you in real-world business situations. Studying the scenarios and completing the activities, with guidance from the chapters, will help you become a more effective business writer. Topics covered include how to

- develop writing strategies that suit your goals, audience, and writing situation (Chapter 2).

- craft effective résumés and cover letters and succeed in other aspects of the job search (Chapter 3).

- design visually appealing, reader-friendly documents (Chapter 4).

- communicate effectively with colleagues within an organization (Chapter 5).

- build and maintain positive relationships with audiences outside an organization (Chapter 6).

- tackle more complex business writing projects, like grant proposals and business plans (Chapter 7).

- avoid some common gaffes in business writing (Chapter 8).

- strengthen your potential by applying the writing skills that strong leaders use (Chapter 9).

CHAPTER 2

An Overview of Business Writing

Understanding the Central Concerns of Business Writing

Chapter 1 introduced you to the challenge of getting *inside* the business writing process. This chapter offers a broader, more conceptual exploration of the factors that affect the practice of business communication.

Seeing the Big Picture

The flowchart on the page that follows captures the central considerations of business writing communications; most of the points about purpose, audience, and best-practice strategies will eventually be second nature to you as you become experienced professionals. As students of business writing, however, it's likely that you don't yet have the experience to follow these guidelines automatically. Before you begin to compose any piece of business writing, you will need to spend some time considering your professional role in the communication chain, your purpose in writing, the audience you must reach, the strategy that is most likely to achieve the desired goal, and the best medium for your

Business Writing Overview

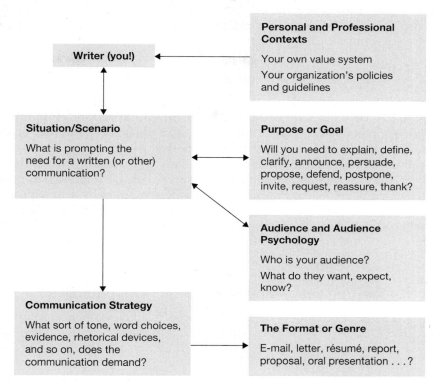

communication purposes. As you craft any communication, you will also need to pay attention to details and avoid mistakes that could make a bad impression on your readers.

Many of these considerations should sound familiar from your previous courses in writing; now you are applying these important skills to real-world business situations.

Starting with *You* as Person, Employee, and Writer

The box at the upper right corner of the flowchart on this page ("Personal and Professional Contexts") suggests a number of things you need to consider, on both a personal and a professional level, to communicate effectively. For example, do your own values for addressing issues and interacting with others coincide with the character of the organization for which you are working? If you and the company are a good fit in these respects, you are fortunate. If you are asked, however, to communicate in ways that don't coincide with your own personal values, what are you going to do about this dissonance? Finding a rewarding balance between the company's purposes and operating principles and your own commitments is worth considering throughout your career and, if possible, even before you commit to a position.

Understand Your Level of Authority

Again in the box labeled "Personal and Professional Contexts," note the importance of your level of authority in the organization. How much authority do you have, for example, to rectify a particular customer problem? Do you need to check with your supervisor before you offer a special discount or guarantee? How much can you actually share about company policies with an inquiring reporter? To what extent, if at all, can you require your colleagues in the next office to abide by a new procedure? Would it be acceptable for you to draft and submit a revised company mission statement on your own initiative, or would you need to get permission from one of the managers or belong to a certain task force in the organization?

Look before You Leap—and Stay Alert for Opportunities

The challenge of balancing personal and professional goals and values is something that every working person needs to explore individually, but there are a few general principles that you can broadly consider as you enter the working world:

- **Look before you leap.** Get to know a bit about the people and the authority structures at your organization before you move into an activity not explicitly delegated to you.

- **At the same time, be prepared to take some chances.** No one gets very far in work (or in life!) without exercising initiative and imagination.

- **Keep alert to new opportunities within your organization.** Selectively volunteer for a group task or an individual project, even if you've never engaged in this task before.

Regarding the last bullet point, remember that very few people are actually trained to do all of the things they encounter on the job. That's certainly true of professional business writers, many of whom have not had the advantage of college courses in business-related writing. So, if you do know the basics of business writing, and your company needs, for example, to write a grant proposal, raise your hand and offer to draft one. Even if you have never written this type of document, you can easily find samples and guidance online and in books devoted to the topic. (You might even read a proposal written previously by someone at your company.) You can also ask for advice from leaders in your organization or from colleagues in your field. Taking the initiative on writing projects and reaching out to others for advice won't just give you valuable experience in new tasks; it will also help you establish valuable connections.

Being alert for and eager to take on work opportunities is a big factor in job success. If you have confidence that you possess good writing and critical-thinking skills and that you can learn and master tasks that you don't yet know, then jump in! Even if you are unable to perfectly pull off a task the first time, your initiative and leadership might be noticed and rewarded with more opportunities in the future.

Familiarize Yourself with Your Organization's Principles and Values

As you can see, your role as a business writer is complex—even before you start composing. *In every instance, you are writing not just for yourself but rather on behalf of the company you represent on a particular question or issue.* Business writing is not an opportunity for personal expression (though your own values are certainly involved in the process). You must be familiar with your organization's guiding principles and values. Is there a company policy manual that you can peruse, perhaps even a style manual for different types of print and digital communications? Can your immediate supervisor outline your areas of responsibility and authority; alert you to sensitive histories within the company; and, in other respects, mentor your early employment experience? Can you review examples of previous communications from the CEO, a customer-service representative, and the director of Human Resources? These samples may not always be the best models for your own writing practices, but they will give you a point of reference. They will show you, among other things, how communication styles are shaped by different levels of organizational authority as well as by individual personality traits.

Keeping Your Purpose in Mind

Whether imagining yourself inside the scenarios in this book or facing them head-on in your first position, you will need to determine exactly what the purpose of your communication is before you start to write. The "Purpose or Goal" box of the diagram on page 21 reminds you of the myriad things you might want to accomplish with different documents. For example, you might want to define or explain a company policy on vacation-day accruals, clarify the resources available to help employees make investment choices for their 401(k) retirement funds, persuade your research team to use a newer technology, defend the company in a public dispute, reassure and apologize to an angry client, thank a group of associates for their ongoing dedication, and so on. (For more information on the various purposes of business writing, see Chapter 1.)

Many of your documents might actually have multiple purposes. For example, you might be supplying financial information on your company while also trying to persuade the recipient to invest in the company. Or you could be delivering disappointing news to a job applicant while also encouraging her or him to apply for future jobs in your company. Whatever the case, you need to know why you are writing in the first place: What do you want the document to accomplish? Do you want to present information as clearly and concisely as possible and make a persuasive case at the same time? Prevent a lawsuit through clarification of facts and convincing reassurance? Persuade colleagues to accept a difficult policy change that benefits the long-term health of the company? Writing with a clear understanding of your goal(s) in view will help you create a well-organized and effective communication.

Understanding Your Audience and Audience Psychology

In every type of business writing, you need to keep the audience in mind. *All* readers want and need you to communicate clearly. That said, many other needs and expectations of an audience vary widely. Here are a few general guidelines that you should consider before writing:

- If the purpose of your document is simply to explain something noncontroversial, you must consider, research, or even directly ask how much your audience already knows about the topic so that you don't, on the one hand, belabor the obvious or, on the other hand, skip over points of information that are not well known to your readers.

- If your purpose is to persuade readers toward a new course of action, you need to anticipate why they might resist your argument and evidence. Anticipate the likely concerns and objections and respond to them systematically so that your colleagues will rethink their original points of contention.

- If you want to mollify an unhappy constituent, you need to consider what prompted the unhappiness, how much company responsibility to acknowledge, and what to do to improve the situation. If you think that a placating measure, such as offering the unhappy party a special discount or future deal, would improve the situation, check with a supervisor before moving forward with it.

- If you must present negative news to your colleagues, you need to determine the levels of their reliance on the resource or opportunity that will be adversely affected, what company or personal histories might lurk in the background, and how you will persuasively justify the necessity of the negative impact.

There are *many* other audience scenarios beyond these, and in every case you will need to choose the tone of voice, types of evidence, and rhetorical strategy that will best meet your communication goals. You will also need to select the most effective medium for addressing the situation. For example, what evidence would you need to convince fellow employees that the reductions in their medical benefits are a financial necessity for the company? What would you write to convince a dignitary or community leader that it's worth her while to be a keynote speaker at your organization's upcoming event? How would you persuade experienced managers that newer technologies will significantly improve the quality of a tried-and-true product?

Determining a Communication Strategy

Now let's turn to the "Communication Strategy" section of the graphic on page 21. Suppose that you now understand the issues surrounding your particular scenario, your goal or purpose in writing, the audience being addressed, and the best medium to use when addressing the members of your audience. Your strategy in composing will depend on these factors and many other variables as well. Underlying all strategies will be the desire to communicate clearly, to offer sufficient evidence and reasons, and to demonstrate that you are informed and fair-minded.

The Importance of Evidence

One critical choice, especially in addressing a controversial topic, is the amount of evidence and information you provide. For example, if you provide too few reasons regarding the upcoming (unpaid) furlough days affecting certain staff members, the staff may view the decision as arbitrary, not as a necessary strategy for the company's financial survival. They will likely assume it was an easy way for the company to save money and will resent the lack of explanation. However, if you provide extensive information on the furlough issue, those opposed to it will have more material on which to base a counterattack. When given too much information, people often seize upon a few pieces of information at the expense of the broader picture. For these reasons, you will need to strike a balance in such situations, sufficiently convincing your constituents of your point of view while not overloading them with information that could provide more fuel for a prolonged dispute.

The Importance of Tone

The tone of the message is extremely important—even a couple of mis-chosen words can undercut the effectiveness of your writing. *Tone* is the writer's attitude toward the topic, toward the audience, or toward both. Tone determines, for example, whether the message sounds friendly or angry, decisive or vacillating, caring or dismissive, supportive or aloof, sensitive or sarcastic, formal or relaxed. Whenever possible in your business writing, embrace a tone that implies your fair-mindedness; professionalism; respect and empathy for others; and commitment to reason, evidence, and a collaborative process. For example, consider these individual word choices and phrases and the different tone conveyed by each:

1. An accusation from a supervisor vs. a firm reminder of a staff member's responsibilities:

   a. *You have done a terrible job keeping track of the expenditures . . .*

   b. *An important part of your work is the tracking of expenditures . . .*

2. A top-down decision vs. a collaborative process among colleagues:

   a. *The work schedule for the holidays is as follows . . .*

   b. *It's time once again to come together to figure out the holiday schedule . . .*

3. An abrupt and thoughtless announcement vs. a tactful and composed statement:

   a. *Jack Snow died this past weekend . . .*

   b. *We have received the sad news that one of your colleagues, Jack Snow . . .*

4. An unapologetic reply vs. an acknowledgment of error followed by a correction:

   a. *Your order is on its way . . .*

   b. *We regret the shipping delay in your last order and are pleased . . .*

5. A negative announcement lacking context and sympathy vs. a compassionate notice with justification:

   a. *Your son has flunked out of Troutbeck College . . .*

   b. *Each term the faculty review committee . . . We are sorry to tell you that your son has not met the continuation standards . . .*

6. An abrupt policy change with no justification vs. a respectful announcement with evidence provided:

a. *All employees must cease immediately from parking in . . .*

b. *Given the serious shortage of parking for our customers, we ask that all staff park in the areas designated as . . .*

As is true in so many communications, business decisions regarding tone are complex and situational. While the latter (b) examples of tone are usually preferable, on some occasions the more abrupt and more decisive tone of the former (a) examples might be necessary. For instance, the supervisor writing in example 1 on the previous page might have good reason to be stern with an assistant who has received prior offers of guidance, and warnings, regarding his or her responsibilities. In most cases, however, you should cultivate a sense of respect and partnership among colleagues and clients by choosing a tone that suggests that you, as colleagues, are all in this together.

Paying Attention to Details

The most effective business writers don't just concern themselves with big-picture issues, such as purpose, audience, and strategy. They also pay attention to details, such as choosing the right words and making sure that their writing is error-free.

Using Clear Vocabulary

Clear, accessible, broadly understood words are central to most business writing. The challenge is, often, to convey complex issues using a plain, ordinary vocabulary. It's rare for even a highly educated CEO or CFO to communicate with a complex vocabulary laced with Latinate terms (for example, "loquacious," "exacerbate," "propitious," or "obfuscate") or with technical terms not understood by a broad audience (for example, such financial investment terms as "Sharpe ratio," "bid-to-cover ratio," and "arbitrage"). In general, technical vocabularies are reserved for specialists writing to one another.

Consider the plain-English word choices made by William Clay Ford Jr., executive chair of Ford Motor Company, in this excerpt from his preface to the *Ford Annual Report* for 2012. The CEO, educated at Princeton University and the Massachusetts Institute of Technology, wants to reach a broad audience of investors, reporters, customers, and others.

A Message from the Executive Chairman

In 2012 Ford Motor Company continued to go further to meet the needs of our customers, the challenges of our industry, and the issues confronting our world. Our efforts produced strong financial results and our fourth year in a row of positive net income.

We expect 2013 to be another strong year for our company. We anticipate our outstanding performance in North America will continue, with higher pre-tax profits than in 2012. We are refreshing our entire product line in South America and continuing to invest for growth in Asia Pacific and Africa. The transformation of our European operations, which is aimed at achieving profitability under difficult economic conditions, is on track and ongoing.

We will continue to focus on producing vehicles with best-in-class quality, fuel efficiency, safety, smart design and value — built on global platforms. They will help us toward our goal of increased global sales and market share, as well as support our ongoing commitment to reducing the environmental impact of our vehicles and operations.

Our strong showing in the electrified vehicle market is a good example of how great products can help build a strong business as well as a better world. In 2012 we introduced six new electrified vehicles in North America, including hybrid, plug-in hybrid and pure battery electric models. By offering a variety of vehicles, we make it easier for customers to embrace fuel-saving technologies.

Unless you are writing to specialists who share your understanding of a particular field of business and its terminology, opt for a clear and easy vocabulary. For example, if you are developing a template for responding to unhappy customers, you would probably *not* write this:

We regret that your order was not expedited properly and that our unhelpful representative exacerbated the problem.

The words "expedited" and "exacerbated" are not mysterious to many people, but they might be misunderstood by some readers. The next example uses a plainer vocabulary:

We try always to provide the most efficient service, and we regret the delays you experienced with your recent order and the unhelpful customer service.

Remember, as you craft business communications, tuck the complex (and certainly enriching) vocabulary away for other thinking and writing tasks, and use specialized terms only for specialist audiences.

Proofreading

Typos, misspellings, grammatical errors, faulty sentence structures, and punctuation ambiguities are not acceptable in professional writing. Even a few such errors in a document can discredit the author and distract the reader from the work's important content. In longer documents a number of errors may cause readers to lose focus as they anticipate the next blunder in the text.

Taking the time to proofread carefully is a critical part of professional writing. No writer, no matter how well informed and meticulous, is immune from errors. During all of your writing you should have a good writing handbook and a dictionary by your side (or bookmarked on the Internet). If possible, have another person proofread your text after you have made your best efforts to correct errors. All of us tend to "normalize" our reading of our own writing: that is, we automatically fill in the blanks (of missing words, for example) and glide right past grammatical and mechanical errors (for example, faulty subject-verb agreement or the distinction between singular and plural possessives) because we know what we meant to say. It's a perceptual and cognitive impasse that even the best writers face daily.

Responding to Real-World Writing Scenarios

Drawing on the "Business Writing Overview" on page 21 and the discussion that follows it, let's see how the concepts of purpose, audience, and strategy apply to the following scenarios.

Analyzing a Writing Scenario: Explaining a Policy Change

The background. For the past decade, Media Inc. has been fairly casual about individual employees deciding when to use their earned vacation days. This informality has led to inadequate office coverage—particularly during holiday periods—and has occasionally delayed important business decisions because key staff members were not available for consultation.

The purpose. The director of Human Resources at Media Inc., Shelley Seidman, has asked the benefits assistant, Amanda Jaffurs, to draft an announcement that clarifies employees' use of accrued vacation days.

Jaffurs has been provided with the content of the policy changes, and the director will review and possibly modify Jaffurs's draft before sending it out under her own name. Jaffurs's central purpose is to construct a clear, accurate, and tactful draft for colleagues who will undoubtedly have very mixed feelings about reduced flexibility in their use of vacation days.

The audience. Even the most dedicated employees value their vacation days greatly, and in recent years at Media Inc., employees have enjoyed being able to choose their own vacation times. Media Inc. has held two open discussions regarding the need for better coordination of vacation time, so employees have already heard the managers' concerns and have had an opportunity to express their views on the subject. Most employees understand the need for a more organized and better-coordinated vacation process, though a few have been vocal in their objection to losing any autonomy.

The communication strategy. The director wants the announcement of the vacation-approval guidelines to be responsive to the sensitivities surrounding this issue, to be friendly and respectful, and to continue to allow for some flexibility in vacation-day choices.

A First-Draft Explanation of the Policy Change

The first quick draft that follows conveys the necessary information concisely, but it does not feature the nuances of tone and context needed to win a relatively positive reception from the staff:

DATE: July 25, 2016   **[DRAFT A]**
TO: All Employees of Media Inc.
FROM: Shelley Seidman, Director of Human Resources
Subject: POLICY CHANGE: Approval Needed for Vacation Days

Pursuant to our recent discussions of the institution of vacation-day approvals, management has decided on the following:

All employees must submit vacation requests to their area director no fewer than three weeks in advance of the desired vacation period. The form for this request is posted on the HR Web site.

We appreciate your cooperation as we institute these changes. Direct any further concerns or questions to seidman@mediainc.com.

Immediately, the writer sets a tone of negativity and inflexibility with the chosen subject line. The capital letters make it seem as though an announcement has been handed down from above. Even the first word in the e-mail, "pursuant," sounds like legal jargon. The overall tone is flat and official. No background reminders regarding the original problem of uncoordinated vacation days are included; the writer should have supplied this context instead of trusting that everyone would recall earlier conversations on the topic. There is no sense in this draft of a colleague writing sympathetically on an issue that has serious implications for the personal lives of everyone in the office—including her own.

A More Detailed Explanation of the Policy Change

After reading the first-draft memo, the director of Human Resources asked Jaffurs, the HR assistant, to write a more fully developed draft, one that offers more context for the issues being addressed and more evidence that the Director appreciates her colleagues' concerns. Here is what Jaffurs came up with:

DATE: July 25, 2016 **[DRAFT B]**
TO: All Employees of Media Inc.
FROM: Shelley Seidman, Director of Human Resources
Subject: Coordination of Vacation Days

Our company-wide discussions on July 8 and 10 regarding the need to coordinate employees' selection of vacation periods were very helpful to crafting a revised policy.

Vacation periods are very important to everyone, and all of us have different individual and family needs regarding periods away from work. We hope the revised policy will preserve an element of flexibility while providing much better coordination of staff availability for crucial company operations.

The new policy, in its basic form, is as follows:

"All employees of Media Inc. should request vacation periods of three workdays or longer at least three weeks in advance of the requested period. Requests should be made on the form available from HR to the employee's area director, who will consider the workloads and responsibilities in his or her area before submitting the request to the director of Human Resources."

The policy also notes, "Whenever possible, the flexibility of employee choices will be preserved. The area director and the director of Human Resources will discuss options with the affected employee(s) before any request is denied or modified."

---

continued

You can find the full text of the policy and the request form online at www.mediaHR /policies.com. I hope the revised policy balances our concerns for workforce coordination with the personal importance of vacation planning. Please don't hesitate to ask me for clarification if needed: ext. 6628, or seidman@mediainc.com, or by appointment in office 2233. Thanks very much for your input and cooperation.

Draft B has the human touch, a quality that nurtures cooperation and respect among colleagues. The subject line suggests a more positive connotation without sugarcoating the general message. The writer thanks her colleagues for their valuable input and shows understanding of their interests in anticipating and planning vacation days under this new policy. She includes the basic tenets of the new policy, identifies the location of relevant materials, then closes with a second thank-you and an invitation to her colleagues to seek further information from her if necessary.

While Draft B will not make everyone like the messenger or happily embrace the new vacation request policy, at the very least it will allow readers to feel that a thoughtful process was conducted, that their voices were heard, and that the company's management is not cold and autocratic. This approach to business communication, repeated over months and years, helps foster a cooperative and productive work environment. The more patient and nuanced the approach (whenever that is possible), the better off the business will be in the long run.

Analyzing a Writing Scenario: Weighing the Costs and Benefits of Conciseness

The previous scenario provides a good opportunity to examine the most common business writing goal: *conciseness*. The traditional argument for conciseness is often that businesspeople are very busy and thus want, as quickly as possible, to glean the main point of a piece of writing. (Consider the colloquial acronym KISS that you often see in business writing manuals: **K**eep **I**t **S**imple, **S**tupid!)

Writing concisely is a virtue in business communications if it means not wasting words—not writing more than is needed to make your case or to convey the necessary information. In certain circumstances, however, conciseness can be counterproductive:

- when it leads to communication that readers might perceive as terse, dismissive, or uncaring

- when it means that the communication is short on evidence or short on the information needed to persuade different audiences or to clarify an issue or procedure

- when it leads to writing that lacks reassurance to, say, a disgruntled customer or to colleagues uncertain about recommended product improvements

A Brief, Formulaic E-mail

We've all received generic e-mail notices like this one:

> Dear Customer:
>
> We appreciate your concerns and will address them as soon as possible . . .

No one is satisfied when he or she receives formulaic messages of this sort. There is no sense that anyone at the company actually cares about the customer's concerns or that anything will be done about them. The message is certainly concise, but it is almost worthless as a business communication.

A More Developed E-mail: Example 1

Now let's add a thin layer of care and precision to the message:

> Dear Mr. Franklin:
>
> Southampton Mutual regrets the late shipment of your order and the defective parts you received. A new shipment is being sent today by FedEx.

This response is a small improvement over the first, in that a company representative seems to know the nature of Mr. Franklin's problem, apologizes for the mishandling of his order, and is taking steps to correct the problem.

A More Developed E-mail: Example 2

An even more developed message, however, one that would solidify customer relationships over time, might sound like this:

Dear Mr. Franklin:

Southampton Mutual has appreciated your business for the past several years and is very concerned that you experienced problems with your last order. We sent a complete replacement order to you by FedEx today and will discount your invoice by 20 percent. The FedEx tracking number is QR67331890.

Again, we regret the recent mishap and the frustration you experienced with our service. We are examining our packaging and shipping processes to be certain this problem does not occur again.

That being said, if you do have remaining questions or encounter a problem in the future, please contact Dawn Russo, our customer relations supervisor, directly at 805-666-9999 or by e-mail at drusso@Southampton.com. We appreciate your patience and look forward to continuing to serve your business interests.

This longer communication has several advantages: the company seems actually to care, has promised to resolve the broader shipping problems, has sent the replacement order (and provided a tracking number), and has given this important customer a discount. The communication is reassuring and respectful throughout and offers specific company contact information for the customer's future use.

Crafting communications with just the right tone and level of detail might take longer, but it also builds stronger business relationships because it suggests that you care about the consumer, customer, or other constituent. You don't need to construct this message from scratch in every situation. Many well-run organizations, small and large, save templates of effective routine communications and modify the specifics to fit a particular scenario. Some also build helpful e-links into their customer communications. (For example, Amazon.com's communications include quick links that let you track your order or contact customer service.) Keep in mind, however, that a communication that sounds too generic will not please a customer who has a serious concern. When time allows, try to individualize a template to address the specific audience situation.

As you respond to business situations in writing, you will be aided not only by templates (or by previous examples of similar documents) but also by increasing familiarity with recurring structural and situational archetypes. For example, the structure of a longer business report typically requires data, anecdotes, and other types of evidence to support the observations presented. It often includes an Executive Summary (a brief digest of the report's key elements), a Background section (a map of the topic's recent history), a clear Problem Statement or description of the current question to be explored, and, often, a set of Recommendations at the end. The titles of these sections and what they contain might vary, but in general, longer reports will feature these same components and other strong similarities across structure.

As you discover these structural commonalities among different writing genres, you will also learn audience strategies that can be applied to related scenarios. For example, a useful tactic for declining a customer request is this underlying message: "I wish that I could, but I can't because . . ." That formula can guide your response to a customer complaint, to a colleague's promotion request, to an associate's request for confidential information, or to a qualified job applicant whom you are unable to hire. The "I wish I could" component makes clear that a "no" will soon follow and simultaneously implies sympathy for the recipient's interests. The "because" provides a reasonable explanation as to why the request can't be fulfilled. Remember that when you give reasons they should be clear and persuasive to your audience, not vague or mysterious. Colleagues generally want to know *why* something important is happening.

An E-mail That Offers Specifics and a Longer, More Detailed Document

Below is another routine communication that is concise but also nuanced with a respectful tone and useful specifics. The writer provides the necessary context regarding an important attachment to the e-mail and mentions that he will be sending a more comprehensive (and corrected) document shortly:

Dear Ms. Sanchez:

Attached is the report you requested. Please note that the financial figures on pages 12–15 are being re-examined, and the revised figures will be forwarded to you by the end of this week. In other respects, the report is complete and should answer many of your questions. If I can be of further assistance before then, don't hesitate to contact me at 672-559-1938, or by e-mail at chilcote@acefinance.com.

Sincerely,

Lee Chilcote
Finance Assistant

The previous examples suggest the virtues and limitations of conciseness in business writing. Certainly, do not waste words by digressing, by repeating points already made, by providing more information than is necessary, or by belaboring the obvious. But do take the time to create documents that present the necessary information in a logical order, that show respect for your audience, and that use the evidence and writing strategies that are most likely to make your case.

Applying What You've Learned

The following activities ask you to apply the broader principles discussed in this chapter. As you craft responses to them, think carefully about the tone you want to use, how you will strengthen (or at least not damage) your company's relationship with the audience, and how you will accomplish your overall purpose.

APPLICATION 2-A

Inquire about Office-Space Needs

The background. In any organization there are resources that many employees would like to see distributed more liberally or fairly—for example, salaries, health benefits, tech support, or office space. In a future job, you may be involved in the process of deciding how such resources should be allocated, perhaps asking fellow employees about their interests and desires. If you do need to gather such information, you must be careful not to raise employees' expectations or to imply any promises about future allocations. Your task instead is to seek colleagues' input in order to aid the decision-making process. The final decision might or might not meet employees' hopes.

In this application, you are the assistant to Voletta Williams, the facilities manager for TrustUs Insurance Company. You have been asked to inquire about current and anticipated office-space needs among a staff of 55 employees, 35 of whom currently work in cubicles and 20 of whom have private, enclosed offices. The company has leased more space in the adjacent building and will be able to expand the square footage in some of the cubicles and provide separate offices to eight to ten of the employees currently in cubicles.

The purpose. Voletta Williams wants you to obtain information directly from the employees regarding their current work spaces so that she can devise a space-allocation plan that makes sense for the daily business activities at TrustUs Insurance Company. She needs practical workload information and evidence from each staff member so that she can present a plan of action to the senior managers. She does not want you to make any implicit promises regarding office facilities; she wants you just to gather information.

The audience. All employees value their individual office spaces, so you can expect a good deal of excitement and anxiety resulting from this inquiry. Consider how you will write in a friendly and reassuring way to your colleagues and at the same time encourage them to make a fact-based case for their individual needs and desires. You don't want to end up with an inflated, impassioned wish list of little use to your boss. At the same time, you also don't want to unnecessarily incite panic among employees regarding the future of their space.

The communication strategy. You will need to provide employees with the basic facts (background) regarding current space distributions and the possibility of enhancements to some of the cubicles and offices. Also let them know about the evidence they should supply if they are currently in a cubicle and want to be considered for larger or enclosed work spaces (for example, a business-related need for privacy, inadequate document storage, communication patterns with other cubicles or offices, or technology needs that affect work space). Be clear that you are gathering the information for Williams and the senior managers and that you must receive the information by a particular deadline. Throughout your memo, cultivate a reassuring tone and a rhetoric that shows your understanding of the issue's importance, but be careful not to make promises or claims that have not yet been confirmed by your superiors.

APPLICATION 2-B

Seek Volunteers for a Nonprofit Organization

The background. Sometimes a company needs employee volunteers to help shape a company policy or process or to plan an event. For example, employees may be asked to join a task force on emergency preparedness, to plan a major company social event, or to reexamine vacation policies. Some of your colleagues will readily volunteer for these above-and-beyond, unremunerated commitments, while others will not want to lend even more time to job-related responsibilities. It's quite an art to bring volunteers on board for the general good of the organization and to make them feel rewarded through the volunteer effort itself.

In this application, you are the coordinator of volunteer outreach for the nonprofit organization Reading Partners, a group working to enhance the reading experiences and literacy of grade-school students. You need to write a letter to the leaders of a dozen local organizations (such as the Elks Club, Kiwanis, and the National Organization for Women) to solicit volunteers to work with your nonprofit's after-school programs for children aged 5 to 12 years old. You are asking the leaders of other organizations to contact their membership for potential volunteers.

The purpose. You need to solicit 25 volunteers to engage in reading activities for children at least twice a week for a total of at least four hours each week per volunteer. Once you receive the names and contact information from the volunteers, your organization will schedule several orientation sessions to outline expectations, goals, and practical logistics. Your group will also train the volunteers on your program's reading curriculum.

The audience. Your audience in this case is already oriented toward community service, but members are also very busy and will need to be persuaded that reading proficiency among young children is a serious concern and that the volunteers' time will be well spent. (Do a bit of Internet research so that you can incorporate some key literacy statistics in your letter.) You should also mention some special perks for those who successfully engage in the volunteer service. (For example, they might receive such benefits as invitations to your fund-raising dinners or special access to the authors you sponsor each month at the local library.) Everyone who has worked with volunteers knows that they are committed to helping, but they also relish special opportunities to reward their free services.

Keep in mind, too, that you are soliciting these volunteers through other organizations. You want to make it easy for the leaders of the other nonprofits you contact to pass along your message to their own members.

The communication strategy. Write in the spirit of one nonprofit organization to another; that is, appeal to the recipients' established interests in civic engagement and community responsibility. Provide enough information (including the childhood literacy evidence referenced previously) about your program to appeal to potential volunteers, and provide clear contact information and the time frame for launching the meetings and training sessions for new volunteers. Make it worth the while of the other nonprofit leaders to pass along the message to their employees and investors. Assure them that, in similar future endeavors, you would be happy to do the same.

Write this piece in the block letter format presented on page 89 of Chapter 4, "Business Document Design, Formats, and Conventions."

APPLICATION 2-C

Coordinate Accounts Receivable Information

The background. Businesses often look for ways to make various processes more efficient; for example, they may seek better coordination among office functions. Instead of just announcing new rules or guidelines for improving efficiency, managers may be better served by first getting input from employees who handle different parts of a process that needs reform. This approach is likely to result in better policies with greater employee support.

The person in charge of gathering such input needs to be candid about the problem or the improvements needed but at the same time not point a finger of blame at those whose feedback is needed for the inquiry to be successful.

In this application, you are the finance assistant for a medium-sized online retail store specializing in household goods. An important part of your job is to track purchase payments, returns and refunds, and payment defaults (for example, because of faulty credit card information). You and your boss have discussed a need for better coordination among some of your colleagues who handle different parts of this information. Your boss, Jerome Kim, has asked you to send a friendly e-mail to these associates to organize a meeting in which attendees will discuss how to improve information flow.

The purpose. Kim also oversees the offices to which you will be sending the e-mail, so his authority stands behind your request, and he will be cc'd on the e-mail. You want to arrange a meeting for the coming week among seven key associates to explore improved links for everyone's benefit. Kim will chair the meeting.

The audience. Let's assume that the colleagues to whom you are writing are entirely competent and have experienced their own frustrations with the current flow of financial information. Thus, they will be pleased to participate in efforts toward improvement. They are also busy with appointments and committee meetings, however, so you will have to coordinate a number of schedules to find a common meeting time and place. You will also need to ensure that your colleagues come to the meeting prepared to share their experiences and suggestions for improvement.

The communication strategy. While you are writing with the sanction and support of your (and your recipients') supervisor, you still want to write in a respectful colleague-to-colleague tone. Emphasize that you are all working with key parts of the larger financial picture; thus, each of you has a stake in wanting to coordinate the overall information flow. Offer several days, times, and a location for the anticipated meeting, encourage scheduling flexibility, and be clear that you need to hear back by a particular time.

CHAPTER 3

Résumés, Cover Letters, and the Job-Search Process

Understanding the Application Process

Whether you're an experienced job seeker or brand-new to the hunt, knowing how to draft an effective résumé and cover letter is key to your success in securing a position and advancing in your field. You need to present your skills, training, accomplishments, interests, and energy clearly and persuasively. You need, in effect, to brag about yourself without sounding boastful: a challenging balance.

Preparing an effective résumé and cover letter will also involve assessing your background and abilities with an eye to identifying every advantage you might have over the dozens (or even hundreds) of other applicants who typically respond to every advertised job opening. But a well-crafted résumé and cover letter aren't the only requirements for success in the job search. You will also need to identify job openings that are the best fit for you and to know how to make a good impression during job interviews. You will benefit, as well, from an understanding of the application-review process. This chapter will guide you through all these aspects of the job search.

You will need to prepare formal applications of this sort not only at the outset of your career but also if you decide to apply to graduate school, to change jobs, or to seek a more advanced position within the same company. The ability to present your talents and achievements persuasively, then, is of ongoing value.

Along with enabling you to get a job and earn a living, the job-search process offers an opportunity for you to reflect on your interests, skills, talents, and the life you hope to lead. Work life occupies enormous amounts of our time and energy over the years—typically, 40 to 50 years for many working professionals. That degree of commitment deserves your close consideration, both while you are seeking employment and periodically during your working years.

Assessing Your Abilities: What Do You Have to Offer?

Your first major job after college may prove to be temporary; ideally, though, you want even that position to be the beginning of a progressive career path that rewards your interests and uses your talents. Every job you have should afford you the opportunity to develop new skills and to make contacts with others in your field. Even if it's not your dream job, you should treat it as an important step in your career journey.

Investigate Campus Resources

To help establish a positive career direction and to decide what to include in your résumé, it's a good idea to create an inventory of your current skills and interests. First, on a personal level, the inventory process will help you clarify the types of work that are likely to be rewarding to you and that draw on the best aspects of your personality traits, skills, and cognitive abilities. Second, the inventory will provide the raw material for your résumé and cover letter.

Your campus career-services office is likely to have self-assessment tools available for you. See, for example, whether you can use the Strong Interest Inventory, the SkillScan inventory, and the Kerwin Values Survey. These are just a few of the assessment tools that can help you articulate your job-related interests, values, and skills. These tools ask a wide range of questions and sometimes link your answers (as in the Strong Interest Inventory) to self-assessments provided by people already working in various fields. Thus, especially with the help of a career counselor, you can see whether your job expectations, skills, and interests seem to fit with the experiences of people working in defined areas. Such instruments may not always be accurate indicators for you, but they will at least stimulate your thinking about your employment values and aspirations. They can also introduce you to job fields and subfields that you never thought about before—or even knew existed.

Also find out whether your campus office subscribes to the annual survey titled *Job Outlook*, produced by the National Association of Colleges and Employers (NACE). The NACE survey is considered the best report on what employers are actually seeking in college graduates. It might surprise you that verbal and written communication skills consistently rank near the very top in the NACE survey—followed by other so-called soft skills, such as the ability to work in a team, solve problems, plan and organize, analyze quantitative data, and research and process information.

Ask Yourself Questions

The good news is that *every* college major and minor uses and enhances these in-demand skills to varying degrees. The following questions will help you figure out precisely what your formal studies can offer a potential employer:

- Have you had the opportunity in your classes to craft many different types of writing (for example, lab reports, analytical essays and reports, persuasive pieces, and proposals)?

- Have you had to present materials orally, whether by yourself or in a team? Did you use visual aids or presentation software such as PowerPoint or Prezi?

- How much teamwork has been required in your classes, and what did you learn about group dynamics and strategies for success?

- Have you had the opportunity to lead a group or to delegate certain tasks to others in a group? What did you learn from these experiences?

- Have you learned how to raise questions about a range of topics and how to find reliable research resources to help articulate answers and resolve problems?

- Have you, especially in your social science and science courses, learned how to analyze and present quantitative information?

Consider your experiences with each of the in-demand skills, both positive and negative, inside and outside the classroom. The more diverse your experiences, the more versatile an employee you will be. Add specific examples to your inventory to demonstrate how you developed these skills and put them to practical use.

Identify Your Research Skills

Think especially about the ways you've learned and used research methods and resources in your courses. What categories of information and ideas have you discovered, for example, with the aid of Academic Search Complete, JSTOR, LexisNexis, and Business Source Complete, or through more specialized data services? How have you used these or other resources to address a problem, define an area of conflict, or support an idea or thesis? This research experience will prove invaluable to your future employers because ongoing research is imperative to keeping current in any field. All businesses that hope to survive and thrive must conduct regular research on their products, services, clients and customers, competitors, and emerging trends. The most effective employees at these organizations must conduct similar research as markets, technologies, and required skill sets change. At the same time, ongoing research will help you anticipate the changes you need to make, or the skills you need to learn, to remain current, and valued, in your professional field.

Consider also what you have learned and experienced by

- taking foreign-language courses, studying abroad, or gaining other international experience.

- taking specialized courses in statistics, accounting, or other areas of quantitative reasoning.

- building technology or software-related skills, through experience with Excel, database software (such as Access or SAP Database), Web design and maintenance, or computer languages.

- participating in a research team, as a research assistant to a particular instructor, or on an individual research project.

- gaining job or internship experience related to your broader career interests.

- gaining leadership experience in a club or organization or on an athletic team.

Each of these bullet points describes skills or abilities that are of great interest to potential employers. Be sure to include these on your résumé and to describe a few key skills and experiences in greater detail in your cover letter.

Note, too, that you might still have time in your college career to take courses in a few of these skill areas if you feel they will benefit you in your job search and future career. If you don't have enough space in a busy academic schedule, consider buying a couple of self-help books that can teach you, for example, how to use Excel or to design and launch a Web site. Sometimes having even basic knowledge in a specific area can mean the difference between your landing a job or being passed over.

Assembling Your Credentials

As you think about and inventory the skills, training, and accomplishments that will make you an attractive candidate for jobs and careers, create a portfolio that brings all of this evidence together in an orderly and attractive format.

While this strategy will be easier when you are more advanced in your career, you can begin now to assemble a portfolio that contains key materials, such as the following:

- letters of reference

- a full description of each major job you have held

- a full description of any internships you completed

- a major piece of research, especially one that is business related (for example, a business plan, a policy document, or an industry analysis)

- a full description of any study-abroad program in which you've participated

- any certificate you have earned attesting to computer or other technical skills

- any record or prize that acknowledges your strong writing abilities, leadership roles, or general academic accomplishments

Once you have a scheduled job interview, submit the portfolio well before the interview day. You can bet that someone at the interview will have your portfolio in hand; moreover, he or she will likely ask you some questions based on the materials you submitted, playing directly to your strengths. The submission of the portfolio also helps distinguish you from less ambitious candidates and gives the interviewers another way of remembering you in a positive light.

As an Internet alternative to a paper portfolio, you might assemble similar material on a personal Web site, making sure the prospective employer receives the Web site link in an e-mail. Constructing a professional-caliber Web site is its own art form and will, undoubtedly, rapidly supplant some of the paper-driven job application processes. (If you have never created a Web site or online portfolio, you might check out Portfolio Builder on Facebook, Squarespace.com, Web.com, Weebly .com, or Wix.com.)

Another Web site strategy, one seldom used thus far, is for you to film and post your own brief "interview" on your Web site. You would need to develop a one- to two-minute script in which you "pitch" your best work-related qualities. You would also need to be sure the Webcam filming is reasonably professional in appearance. Many employers would very likely take a peek at your online interview as a convenient way of saving time and money during the initial screening of candidates. If you can present yourself effectively in this format, you will have an edge for gaining an in-person interview.

Finding Open Positions

There are a number of traditional resources for finding job openings: your local newspaper; general job-search Web sites, such as Indeed.com, JustJobs.com, SimplyHired.com, and craigslist.org; and niche job sites for your particular field. However, many companies never post job openings to the public; instead, they rely primarily on internal referrals from current employees.

The following sections will help you make the most of both inside-track and traditional resources for finding job openings.

Unadvertised Opportunities

To seek out these unadvertised opportunities, it's important to tap your network of peers, instructors, friends, and family. You can also speak with the counselors at your campus career-services office. Especially consider the following strategies:

- Talk with your teachers, who might be aware of an opportunity related to your field of study or among the staff positions in the department.

- Attend career fairs on your campus or in your larger community. Talk with representatives at these gatherings, and be sure they know your name and interests. Collect business cards and write to the representatives a week or so after the fair to express your interest — and be sure to include your résumé.

- Attend recruiting sessions on your campus. Typically, the career-services office hosts such visits, and staff there will be glad to share the schedule with you and describe the process.

- Ask employed family members and friends what they are hearing about current or emerging job opportunities at their own companies.

- If you currently have a job or internship, ask your supervisor and colleagues about possible openings at the company.

- Request informational interviews at companies that do work that is related to your skills and interests (and at which you might like one day to be employed).

Your ultimate goal is to be creative and proactive in seeking job opportunities and in making people familiar with your credentials and abilities. Competition for good jobs is very likely to remain high throughout your lifetime, and the slow recovery of the world's economies since the 2008 recession presents an even more challenging situation for college and university graduates. The job outlook is improving significantly, but you will need to be creative and persistent in your efforts to present yourself to potential employers.

Advertised Openings

Check first with the career-services office on your campus: the staff there can familiarize you with any job-search links or services you can access both before and after you graduate. Some of these links are free and others are subscription services paid for by your tuition and fees, so do your research and get your money's worth as a student.

When you explore advertised options, follow these strategies:

- Find online postings related to particular professional areas. Your career-services office can tell you how to access job-search sites that are best suited to your particular field or discipline.

- Don't overlook job openings on your campus, especially as transitional employment or a bridge to other career paths. These can typically be found online at your college or university's employment Web page.

- Check the city, county, and state jobs that may be advertised on local government Web sites.

- Look every week (usually on the weekends) in local newspapers for advertised positions or in newspapers published in a city where you hope to work.

Dos and Don'ts

Be sure to keep a formal record of all the contacts you make, of the advertised jobs of interest to you, and of those jobs for which you apply. For instance, if you made a phone call inquiring about the status of a particular job application, make a note of the date and the person with whom you spoke. If you mailed or hand-delivered a paper application or submitted materials online, keep a full record of what was submitted, to whom, and when. A successful job search is a job in its own right, one that requires your best managerial and organizational skills. You need to know when to follow up with a potential employer and how to do so effectively.

For instance, it's a potentially serious error to once again phone a manager with whom you spoke just yesterday or to send an e-mail that repeats what you already communicated. The manager might get the impression that you've already forgotten the conversation and eliminate you as a candidate, assuming you'll be equally disorganized and forgetful as an employee. Keeping close track of the status of your multipronged job-search process will help you remain visible and determined without becoming obnoxious to employers.

Improving Your Odds in the Application-Review Process

What really happens when organizations review the dozens or even hundreds of application materials they receive for available jobs and internships? Understanding this process from the *inside* allows you to better prepare your materials for review.

Surviving the First Cut

The initial review of applications, whether made by one individual or a small group, occurs very rapidly. Often this process first occurs in a company's human resources department—you should keep in mind, then, that your résumé is unlikely to be viewed by someone in the department in which you wish to work until it has passed this initial stage. The initial reviewer might devote no more than 20 or 30 seconds to a cover letter and résumé—just a quick glance to see if the applicant is in the ballpark of what's required for the job opening. Thus, your application must grab immediate attention with its solid substance, relevant education and experience, and attractive document design. (For a detailed discussion of how to craft an effective résumé and cover letter, see pages 50–68. For more on document design, see Chapter 4.)

In a common first-cut process, the recruiters might sort applications quickly into "yes," "no," and "maybe" piles; sometimes a numerical rating (1–5, for example) is applied instead. Some larger organizations with online application systems use computer software to seek terms and credentials in candidates' materials that are pertinent to the specific job opening. For example, an accounting position might search for "CPA" or for such related terms as "financial," "spreadsheet," and "audit."

Whatever the system used, the first steps of the review can be quick and ruthless. Only a few candidates from the "yes" group will actually be interviewed; the "no" group is out of the running; and it is rare for a "maybe" application to eventually migrate to the "yes" pile.

Increasing Your Chances of Success

Knowing how the application-review process works will help you prepare a résumé and cover letter that will find their way to the "yes" pile. Here are some key steps to making that happen:

- Make sure that there is a reasonable fit between your credentials and the advertised job criteria. Read each job posting carefully to get a sense of the company culture and what each organization is looking for in a candidate, and look for positions for which you feel your expertise is a good fit.

- Be certain that this fit is explicit in the vocabulary and evidence you present in the cover letter and résumé. Tailor your cover letter to each position to which you apply. Stretching your reach a bit is fine, but don't be unrealistic about the possibilities.

- Sell your qualifications strategically, whatever your major. Job candidates who have degrees in the STEM disciplines (science, technology, engineering, and mathematics) are in great demand, as are graduates in such fields as accounting and actuarial science. But even job candidates with these skills will need to make sure that they carefully tailor their application to each position they seek. Good job opportunities are also available for humanities and social-science graduates, though on-campus recruitment efforts for graduates in these areas are likely to be less active than those for STEM graduates. If you are in the humanities, the social sciences, or the arts, make sure that your résumé and cover letter play up the skills called for in open jobs. (For an example of a well-crafted résumé and cover letter from a liberal-arts student, see pages 69–71.)

- Because your goal is to land an interview, be certain that your cover letter grabs *immediate* attention and that all of your materials are designed clearly and attractively. At all costs, avoid careless errors in grammar and punctuation, as these are easy excuses to eliminate candidates. (Winning strategies will be presented in the next section of this chapter.)

- Take into consideration that, despite your good credentials, an element of luck is involved in the hiring process. Each reviewer brings his or her own biases to a review of your cover letter and résumé, and his or her own time constraints to the process of reviewing stacks of applications.

- Submit several applications simultaneously to a number of promising job openings. You will usually need to place your credentials in the hands of five to ten potential employers in order to receive a positive response from a few of them, and you can't waste time by applying just to one position at a time.

- Try not to take any rejections personally. Your not getting an interview may actually have little connection to your qualifications. There's a big element of luck in the application-review process, because it takes the right person happening to peruse your materials at the right time.

- If a hiring firm is relatively nearby, consider delivering your materials in person. You might try to say "hello" to the hiring manager, but even if you don't get past the front desk, the person you meet has now connected your face, commitment, and energy to your application materials. You are no longer entirely anonymous, and he or she might mention a favorable impression to the hiring manager.

- If you are applying online for a position, you may receive an automated response providing you with instructions going forward. Follow these instructions closely. If you are given contact information, follow up within a reasonable time period, but remember that reviewers receive numerous applications each day. They will not appreciate your following up before they've even had a chance to review your résumé.

In general, you should put your best self forward in all aspects of the job application process. In all of your written materials and direct interactions with potential employers, you must try to demonstrate that you are skilled in the areas needed by the employer, conscientious, energetic, and creative.

Next, we will examine two components of job applications in greater depth: résumés and cover letters.

Responding to Real-World Writing Scenarios

In the following sections, we'll examine how two job applicants responded to the challenge of developing a successful résumé and cover letter. As you craft such documents yourself, understand that doing so thoughtfully will not only improve your chances of getting a job in the immediate future but also offer you an opportunity to reflect on what you have already learned and accomplished at this point in your life (your current qualifications); what sort of work you hope to be doing in the future; and what other kinds of education, training, or experience you might need in order to get there. Thus, this process can be a great way to think specifically about your career aspirations and about what you might need to do now to reach those goals.

Analyzing a Writing Scenario:
Crafting an Effective Résumé

Even though an employer is likely to look first at your cover letter, the résumé is especially convenient for those reviewing job candidates. It provides a focused, matter-of-fact summary of your educational background, work experience, and special skills. In contrast, the cover letter allows you to make a case for why your particular interests and qualifications make you a good fit for the job that you are seeking.

Because you probably have written a résumé in the past, you may already know that this document offers a succinct, well-organized, easy-to-read overview of who you are in relation to the world of jobs and careers. In some employment processes, only the résumé is requested (or even allowed), and it is the document most likely to be in the hands of those who interview you.

It is often said that a résumé should be no longer than one page. For most undergraduates, who have comparatively limited job histories and formal credentials, the one-page résumé makes practical sense (for you and for the reviewers). But some undergraduates have fairly extensive and rich work-related experience to present, and in such cases you should let the résumé grow beyond the single page—perhaps even as long as two pages. Let the substance and extent of your qualifications dictate the length of the résumé.

A Problematic Résumé

The résumé should be clear, uncluttered, and visually attractive, and it must represent your talents in sufficient detail. The Jason Stornk résumé on the page that follows does not meet those criteria. It is disorganized, lacks detail, and would almost certainly be discarded immediately by a human resources representative or other initial reviewer.

The overall design. At first glance, any reviewer will notice the glaring inconsistencies in format and design—the varied type styles and formats (all caps, caps and lowercase, italics, and so on), as well as the careless organization of content. In fact, Jason has not taken the time to ensure that *any* of the content in his résumé is aligned correctly, rendering it confusing for potential reviewers. The result is an unattractive document that suggests the candidate is careless, not conscientious about the quality of his work. All of these concerns are aspects of document design, discussed in Chapter 4. In the case of Jason's résumé, the poor design features detract from the content he hopes his reviewers will notice.

The header. Notice that Jason's name and contact information are off kilter and squeezed into the center of the page, not making good use of the full space available. His e-mail address, steelersfan453@gmail.com, is

Jason L. Stornk
1322 Bridge Highway
Alfonso, PA 22197
453-777-9090
steelersfan453@gmail.com

Career Objective: A challenging position that will enable me to contribute to organizational goals while offering an opportunity for growth and advancement.

Education:

Southwest High School   Alfonso, PA
Graduated with Honors

UCSB   Santa Barbara, California
B.A. 2016
Major: Economics, with emphasis in Accounting
GPA: 3.00; Major GPA 3.58
Relevant Courses:
    Economics 101, 117A, 135, 137A-B, 138A-B
    Writing 50 and 109EC
    Computer Science 50 and 60

Work Experience:

Sept. 2015–   Worked in the campus bookstore
present

June 2014–   Internship at Cerebral Accounting Services   Alfonso, PA
August 2015

May 2010–   Managing paper routes for *The Alfonso Ledger*   Alfonso, PA
May 2012

SKILLS:

Great attention to detail
Microsoft Word, Exsell, Power Point, Publisher
Languages: I speak Spanish

REFERENCES:

    Available upon request

okay for friends and family but implies that he does not take himself seriously—and prospective employers won't either. This address should be changed to a more professional one for all business purposes.

Career objective. The career objective in Jason Stornk's résumé is so generic that it isn't worth including at all. His vague generalities about a "challenging position" and an "opportunity for growth" could be said by any candidate about any desirable job. However, you don't want to make the career objective so narrow that it says merely, in effect, "I want to work in your company." Career objectives are useful on résumés only when they suggest your suitability for the job(s) to which you are applying. An effective statement of career objectives will balance your immediate and longer-term aspirations in relation to the goals of the company at which you hope to work. (The improved résumé that follows attempts to strike this tricky balance.)

Education. Jason's educational information is jumbled and vague. For example, what reviewer would recognize the course numbers listed or take the time to look them up in an online catalog? Is Jason's 3.00 GPA, while respectable, strong enough to mention here?

Work experience. The work-experience section is especially vague. Jason offers some intriguing hints here, but what did he actually *do* in his accounting internship? In what respects did he manage paper routes? Unless human resources representatives can get a sense of Jason's specific duties in each of these positions, they'll move along to other candidates.

Skills. Jason's skills section includes misspellings—which immediately belie his claim for "great attention to detail"—and lacks key information. For example, reviewers will have no idea how fluent he might be in Spanish, nor will they understand the extent of his expertise in the software he has listed. Finally, you may have noticed that Jason misspelled "Excel" and "PowerPoint." As a result of all these problems with his résumé, he will almost certainly be passed over in favor of other candidates.

Revisions to the Résumé

Let's tackle Jason Stornk's poorly conceived résumé, addressing his problems with content, organization, and design as we reconstruct the résumé piece by piece.

The header. Jason's original header was off center and cluttered. Here is a more attractive header that contains all of the same information without appearing overcrowded:

Jason L. Stornk
1322 Bridge Highway
Alfonso, Pennsylvania 22197

453-777-9090   jstornk@gmail.com

Note how this header makes better use of the space available and presents the contact information more clearly. A simple design element, the three-layered underscore, has been added to clearly distinguish the heading from the body of Jason's résumé. Notice, too, that the unprofessional e-mail address (steelersfan453@gmail.com) has been replaced by a more suitable address.

On a related point, be certain also to replace any obnoxious or frivolous voice-mail messages on your cell phone or landline. In addition, bearing in mind that increasing numbers of employers investigate candidates online and eliminate those who present themselves unprofessionally, familiarize yourself with the privacy settings for each social-media account you hold, and ensure that any photos, videos, or posts that you would prefer a future employer did not see are kept private or are expunged. (For more on social media pitfalls and guidelines in the job search, see pages 234–40 in Chapter 8, "Business Writing Gaffes in the Real World.")

Career objective. Jason's current career objective is a vague throwaway. Equally worthless would be an objective like this: "An entry-level position in accounting at Asteroid Avoidance Inc.," which merely restates that Jason wants the job to which he is applying. As suggested earlier, if you want to include a career objective, have it suggest that your personal goals line up with those of the company.

Career Objective: To increase my accounting skills in a dynamic work environment, eventually to move into financial management.

Education. If you have significant work or internship experience, that experience will be a powerful aspect of your job qualifications. For most of you, however, at this point in your life your university education is your central, most compelling credential and must be considered carefully and presented well. Here is an improved version of Jason's educational background:

Education:

University of California, Santa Barbara   Santa Barbara, California
    BA expected June 2016
    Major: Economics, with emphasis in Accounting
    Major GPA: 3.58 (on a 4-point scale)

    Relevant Courses:
        Intermediate Microeconomic Theory, Law and Economics, Monetary
        Economics, Managerial Accounting, Income Taxation, Writing and the
        Research Process, Writing for Business and Economics, Computer
        Programming Project, Introduction to C, C++, and UNIX

Universidad de Salamanca   Salamanca, Spain
    Fall 2014 study abroad

    Relevant Courses:
        Intermediate Spanish, Spanish History, The European Union

Southwest High School   Alfonso, Pennsylvania
Graduated with Honors, June 2012

This revised education section establishes consistent margins for each place of education, creating a parallel visual effect. Locations and dates have been clarified, and the names of job-relevant courses are now listed, including Jason's important study-abroad experience in Spain. (As your career progresses, you will continually add to and edit your résumé. It's likely, then, that you'll eventually remove earlier educational experiences like high school to make room for your more current job skills. When you are just starting your career, however, it's fine to include this educational credential.)

Jason has now opted to list only his GPA in the economics major because it's more impressive than his *general* GPA. Please note, however, that he uses the description "Major GPA." To suggest that the 3.58 figure is his general GPA would mislead potential employers. In preparing your résumé, you must decide whether to list a GPA. Certainly a high GPA will suggest that you are smart and self-disciplined, have good work habits, and can meet deadlines. However, since a 3.00 GPA (or even higher) is about average these days at most U.S. colleges, you might want to list only a GPA above 3.00. (As you gain more career experience, you will likely drop all GPA references from your résumé; your new credentials will demonstrate your accomplishments in a professional arena.) Be certain also to list any academic honors, such as dean's list, or any departmental prizes or other special distinctions.

Work experience. The next section of the résumé is one of the most important and often presents challenges to students. If you are a student, you are at a very early stage in your employment history, and you might believe that the jobs that you have held so far are of little substance or relevance to your current career pursuits. You could be concerned that this section will be sparse because of a lack of real-world work experience. You need, however, to carefully consider the work-related and personal skills that you brought to, or developed during, your job as a waitperson, a cashier at the bookstore, a retail clerk, or a camp counselor. You don't want to make inflated claims that might embarrass you during an interview, but you do need to tease out the underlying abilities that made you good at your past jobs and that will make you an attractive candidate for future employment. Keep in mind, too, that your employers have also been where you are—they will recognize the personal qualities that these jobs require and the skills that they build. Here is an improved version of Jason's work summary:

Work Experience:

September 2015–present   Cashier, UC Santa Barbara Bookstore, Santa Barbara, CA
- Mastered all cash-register processes
- Greet customers
- Reconcile register accounts at the end of each shift

June 2014–August 2015   Intern, Cerebral Accounting Services, Alfonso, PA
- Helped chief accountant tally accounts receivable
- Answered routine financial questions for customers
- Prepared accounting information for reporting of state and federal taxes

May 2010–May 2012   Manager of paper routes, *The Alfonso Ledger*, Alfonso, PA
- Organized the routes of 11 newspaper delivery people
- Answered customer calls about missing papers or late deliveries
- Dispatched new deliveries as needed
- Helped interview new delivery applicants

Now the skills and care required for Jason to be an effective cashier, intern, and paper-route manager are clearly outlined, and recruiters will be able to see how these job skills can be applied to their organizations. You want readers of your résumé to say, in effect, "Yes, that job does involve a complex process, patience, and attention to detail." The internship listed now makes clear that Jason has had significant accounting and customer-relations responsibilities. He wasn't just filing papers and answering the phone. The "managing" of paper routes is now clear as well: it required good organizational and managerial skills and the ability to resolve problems and appease unhappy customers.

The improved work-experience section also illustrates parallel structure in writing: each verb in the job descriptions is active and in the past tense, unless it describes duties performed in a current position. Parallel structure can refer not only to grammatical or syntactic consistency but also to the aspects of visual layout we discussed earlier. In each case, "parallel structure" means consistency in style and format. It is up to you how you choose to describe your individual work experiences. You can use phrases, full sentences, or even a more narrative style. You should also select an effective design style, using bullet points or indentations to set off different jobs and job skills. But once you have made such decisions for yourself and for your readers, follow the same, parallel format throughout the section. Also, make sure that the design you use for your work experience is consistent with the design you use for the rest of your résumé.

Skills. The skills section on Jason's résumé was careless and vague. Here is an improved version:

Skills:

Great attention to detail, as demonstrated through my accounting internship
Very proficient in Microsoft Word and Excel; some experience with PowerPoint
    and Publisher
Fluent in spoken and written English and Spanish

In this revision, the claim of "attention to detail" is now supported by the experience Jason obtained through his accounting internship. (He has also spelled "Excel" and "PowerPoint" correctly.) As will be discussed later, in the section on cover letters (see pages 60–68), it is important in job application materials to avoid broad claims about your personal skills and character traits. Rather, you need to support claims with evidence, to demonstrate the existence of these traits through your work experience and duties. Be certain as well to indicate your level of proficiency in the skills you reference, as Jason has now done with his software and foreign-language expertise.

References. Finally, the references section—the standard conclusion to a résumé—is usually a waste of valuable space.

REFERENCES:
    Available upon request

No employer will hire a person into a significant position—the type that you will be seeking at this point in your career—without asking for at least one reference. Employers assume that these references are people willing to write a letter on your behalf or respond to a phone inquiry about you with praise. It is important, then, during your undergraduate years, that you make yourself known to a few teachers or supervisors who can speak well of your talents and commitments when it's time for you to submit job applications. Provide contact information for these individuals only when requested to do so by a potential employer, and only after first asking permission from each reference. To list these individuals in a separate section will only take valuable space you could otherwise use to elaborate on your skills, education, and experience.

However, if your references are well known in the field that you are seeking to enter or are valued associates in the organization to which you are applying, it's a good idea to list them directly on your résumé or to mention your connection to them in your cover letter. For example, if one of your instructors is well known in the field of environmental studies, this would be an impressive reference to list if you were applying to an environmental organization. Or if your former internship supervisor is a manager at the company to which you are applying, you should list this reference as well. If you decide to include references in your résumé, give the full name, title, and contact information for each person.

Finally, if you do have in hand a very strong letter of recommendation, you should not hesitate to submit it with your cover letter and résumé. Even if a job posting stipulates that no letters should be submitted, I encourage you to ignore this caution. The fact is that the inclusion of a glowing recommendation can quickly distinguish your paperwork from a pile of more anonymous materials. The worst that can happen is that a human resources director might withhold the reference letter and submit only cover letters and résumés for further review by an individual or committee.

The Final Product

Once the improved sections are reincorporated into Jason Stornk's résumé, the final product is clear, detailed, and well organized:

Jason L. Stornk
1322 Bridge Highway
Alfonso, Pennsylvania 22197

453-777-9090   jstornk@gmail.com

Career Objective: To increase my accounting skills in a dynamic work environment, eventually to move into financial management.

Education:

University of California, Santa Barbara   Santa Barbara, California
   BA expected June 2016
   Major: Economics, with emphasis in Accounting
   Major GPA: 3.58 (on a 4-point scale)

   Relevant Courses:
      Intermediate Microeconomic Theory, Law and Economics, Monetary
      Economics, Managerial Accounting, Income Taxation, Writing and the
      Research Process, Writing for Business and Economics, Computer
      Programming Project, Introduction to C, C++, and UNIX

Universidad de Salamanca   Salamanca, Spain
   Fall 2014 study abroad

   Relevant Courses:
      Intermediate Spanish, Spanish History, The European Union

Southwest High School   Alfonso, Pennsylvania
Graduated with Honors, June 2012

Work Experience:

September 2015–present   Cashier, UC Santa Barbara Bookstore, Santa Barbara, CA
                           • Mastered all cash-register processes
                           • Greet customers
                           • Reconcile register accounts at the end of each shift

June 2014–August 2015   Intern, Cerebral Accounting Services, Alfonso, PA
                           • Helped chief accountant tally accounts receivable
                           • Answered routine financial questions for customers
                           • Prepared accounting information for reporting of state
                             and federal taxes

May 2010–May 2012   Manager of paper routes, *The Alfonso Ledger*, Alfonso, PA
                           • Organized the routes of 11 newspaper delivery people
                           • Answered customer calls about missing papers or late
                             deliveries
                           • Dispatched new deliveries as needed
                           • Helped interview new delivery applicants

Skills:

Great attention to detail, as demonstrated through my accounting internship
Very proficient in Microsoft Word and Excel; some experience with PowerPoint
   and Publisher
Fluent in spoken and written English and Spanish

Before moving on to the challenges of writing a persuasive cover letter, let's review the criteria that apply to all good résumés.

Résumé Tips

Give yourself the space you need. While a one-page résumé will suffice for most undergraduates, let your individual qualifications dictate the length of your résumé. Some undergraduates have a lot of technical training, numerous awards for their accomplishments, dazzling leadership positions in university organizations or on the athletic field, relevant military training, or even internships or jobs in their discipline. A desire for conciseness and efficiency should not squeeze out such relevant credentials.

If an employer or recruiter actually insists that you submit only one page, follow that direction. Otherwise, let the form accommodate to the function and use the space you need to represent yourself effectively.

Create a clear and attractive document. Your résumé should be easy for readers to follow and should demonstrate visually the care you take with your work. A sloppy résumé suggests carelessness and undermines any claims you make about your attention to detail or conscientious work habits. Especially consider these design components:

- Develop clear headings that orient the reader to what comes next.

- Select a font that looks professional and is easy to read.

Use a consistent font style and size for the linked sections of the résumé. For example, if the "Education" header is in bold, 14-point type, later headers of the same level (such as "Work Experience" and "Skills") should also be in bold, 14-point type. In addition, use parallel structure in your design as you align, top to bottom, the various headings, subheadings, and columns of information. (You should also use parallel grammatical structures as you detail your work experiences and skills—for example, use past-tense verbs such as "mastered," "created," and "organized" when referring to duties from a former position.)

Be sure that any "goals" or "career objective" statements serve a real purpose. Most such statements are wasted, vague generalities, or they are too narrowly geared to the specific job being offered. To be of any use, a statement of this type must marry your short- and long-term career aspirations with those of the company to which you are applying. Otherwise, the company is unlikely to see how your personal goals align with theirs and will move on to other candidates.

Don't sell past jobs short. Even if you regard the jobs you have held as unimpressive, describe your duties in each position in enough detail to reveal the skills you obtained and sharpened through even the most routine tasks. Look for the underlying professional or personal skills required for the jobs you have held: for example, interacting with challenging customers, learning complex cash-register processes, overseeing a work crew, organizing information, doing research, and so on. These accurate and persuasive job descriptors are a key part of a convincing résumé.

Include the following skills, evidence, and experience, when applicable:

- relevant computer and technical skills (especially with Word, Excel, PowerPoint, data-management software, Web site design, social media, and any Web sites or databases used specifically in your discipline or field)

- evidence of your proficiency as a writer

- evidence of your oral communication skills

- evidence of your management or leadership experience, as well as any experience working as part of a team

- international experience (for example, study abroad) and foreign-language proficiencies (including the level of ability in each language)

- examples of your experience with Internet research or with other types of research (evidence that you know how to figure things out in a systematic way, to define problems, and to seek solutions)

Depending on the job to which you are applying, some of these key skills should also make their way into your cover letter. You will learn more about how to write an effective cover letter to accompany your résumé in the next section.

Analyzing a Writing Scenario: Crafting an Effective Cover Letter

While the résumé is a relatively straightforward part of a job application, the cover letter gives you greater freedom to write much more persuasively about your qualifications and interests. The cover letter should resonate with your energy and personality while also including key evidence of your career commitments, skills, and accomplishments. In a sense, it requires you to brag about yourself without seeming boastful!

Essentially, the cover letter is meant to personalize and expand on a few key facts presented in your résumé. The further goal is to convince the reviewer to turn the page in order to see the résumé's more detailed descriptions of your qualifications. Thus, the cover letter cannot be a generic expression of interest in the job. It must quickly gain the reviewers' attention, convincing them to read beyond the first paragraph and to consider how you would fit into their organization.

The ultimate goal of both the cover letter and the résumé is to obtain an interview in which you can expand on your qualifications and directly convey aspects of your personality, energy, and commitment. Getting to the interview stage is key, and seldom will a company hire a candidate without at least a phone or Skype interview.

A Problematic Cover Letter

Let's start with a fictitious cover letter that is bland, nonspecific, and unlikely to persuade the reviewer to read beyond the opening sentences:

September 12, 2016

Julie Liu
Aces Accounting Services
2455 Milpas Street
Cleveland, OH 44101

Dear Ms. Liu:

I want to apply for the job you are advertising. Aces Accounting Services is known for its progressive work environment. I would love to learn and grow in a firm of this sort. I am very detail oriented and have strong people skills.

My background includes a summer job as a bookkeeper's assistant in Cleveland, a job with many responsibilities. I also greeted clients, answered the phone, and did filing. Especially interesting was the work I did on the monthly accounts for several local law firms.

I would be available for an interview at any time. My schedule at Cleveland State is pretty open on Tuesday and Thursday afternoons and Friday mornings. Please let me show you what I can contribute to Aces Accounting.

Sincerely,

Eva Winslow

There is no energy and little specificity in this letter. After reading it, the hiring manager, Julie Liu, will have little idea of who Eva is as a person and will not be sure what skills Eva can offer as a candidate. Specifically, Liu will have the following questions:

- To what advertised position is Eva applying? Aces Accounting might have posted several jobs.

- Did Eva have more significant responsibilities in her internship beyond clerical tasks? There is a hint of this in the sentence about handling "monthly accounts" but nothing specific to suggest actual bookkeeping or accounting experience.

- What evidence is there that Eva has the people skills and attention to detail that she claims in the opening paragraph? Claims must be supported by evidence if they are to carry any weight in a cover letter or résumé.

This letter has no persuasive force or compelling detail to distinguish this applicant from all the others. As a result, reviewers might assume that Eva has crafted a generic cover letter to send to employers en masse. Julie Liu is unlikely to want to hire a candidate who has put no extra effort or energy into landing a job—maybe that candidate will show the same apathy once hired.

Revisions to the Cover Letter

Let's view a revision of this letter—one that develops a much sharper focus—and then go over the process of improving it step by step:

1213 Sheffield Road
Cleveland Heights, OH 44112

September 12, 2016

Ms. Julie Liu, Human Resources Manager
Aces Accounting Services
2455 Milpas Street
Cleveland, OH 44101

Dear Ms. Liu:

Recently I saw your posting for a junior staff accountant on craigslist (Job Order Number: 01260-109545). Your firm interests me greatly because Aces Accounting has a reputation for encouraging its employees to develop their skills and talents. I would bring to Aces Accounting not only experience from a previous accounting internship but also formal training from my Economics and Business major at Cleveland State University.

In the summer of 2014, I worked as a bookkeeper's assistant for Genesis Financial Services, a job that introduced me to many of the areas listed in your advertisement: I helped the head accountant review the general ledgers and correct journal entries, performed rudimentary account analysis, and reviewed bank statements from several different companies. I also honed my skills for interacting in person, over the phone, and through e-mail with a variety of clients.

As my attached résumé makes clear, my major course work at Cleveland State has included several beginning and intermediate accounting courses, advanced courses in financial management, and an enriched context of microeconomics and macroeconomics courses. This formal training in financial skills and economic theory was put to the test during my summer internship and has provided me with a solid foundation for my bookkeeping work thus far and for the new challenge of managing accounts at Aces Accounting.

I will be graduating this June with a bachelor's degree in Economics, at which time I would be available to work full-time in accounting; until then, I could work for you as many as 20 hours each week. I would bring a good knowledge base, discipline, and a strong work ethic to Aces Accounting, and I hope you will decide to interview me. It would be easiest to reach me on my cell phone (216-666-7788) or by e-mail at ewinslow@gmail.com. Thank you for considering my application.

Sincerely,

Eva Winslow

The letter format. Notice that the revised letter uses the full block letter format (see the template on page 89 in Chapter 4). A formal letter of this sort includes the sender's address (on Sheffield Road); the date; the name, title, and address of the letter's recipient (Human Resources Manager); and a professional salutation (Dear Ms. Liu:):

1213 Sheffield Road
Cleveland Heights, OH 44112

September 12, 2016

Ms. Julie Liu, Human Resources Manager
Aces Accounting Services
2455 Milpas Street
Cleveland, Ohio 44101

Dear Ms. Liu:

Even though many companies post job openings anonymously to prevent e-mail and phone inquiries, you should always try to find out the name and address of the company and, if possible, contact information for an appropriate recipient (such as the director of human resources or the manager of the appropriate division). Do a little online research. Sometimes the posting includes the company's name, mission statement, or motto. Sometimes you can Google an unusual phrase in the posting and find its connection to a particular company site. The main point is to create, if possible, a more direct connection between you and the potential employer—this increases the likelihood that your application materials will be seen by the key people.

If you are unable to locate contact information, avoid addressing your letter "To whom it may concern" or to the equally outdated "Dear Sir/Madam." Try substitute salutations such as these:

- Dear Hiring Manager:

- Dear Director of Human Resources:

- Dear Employment Review Committee:

The opening. In the opening paragraph of your cover letter, you must say something significant and attention-grabbing about your credentials if you expect the reviewer to continue on to the second paragraph. Take a look at Eva's approach:

> Recently I saw your posting for a junior staff accountant on craigslist (Job Order Number: 01260-109545). Your firm interests me greatly because Aces Accounting has a reputation for encouraging its employees to develop their skills and talents. I would bring to Aces Accounting not only experience from a previous accounting internship at Genesis Financial but also formal training from my Economics and Business major at Cleveland State University.

In the revised opening above, the writer first identifies the position to which she is applying and includes the job-posting number for reference. This information will help reviewers recognize at a glance the position that the candidate is interested in and whether she is qualified. Eva then mentions what she admires about the company and why she's decided to apply. If you decide to use this strategy in your own cover letter, do so only briefly (perhaps one sentence) and sincerely. Research the company to which you are applying so that you can accurately characterize its goals and reputation. Be genuine in expressing your interest in the company's mission; you don't want to appear insincere or fawning.

By far the most compelling part of Eva's opening paragraph is her referencing her university studies (in this case, economics and business) and her relevant work experience in an accounting internship. This combination of academic training and actual work experience, the linking of the theoretical and the practical, will be one of the most attractive features to hiring managers reviewing applications from university students.

Notice at the outset and throughout the letter that Eva stresses the abilities and experiences she can contribute to the company, not what she hopes to gain from the company for her own personal and professional growth. In an internship application she might balance these two aspects of giving and receiving, but in a job application she emphasizes how she can lend valuable support to the company's initiatives.

The middle paragraphs. The middle sections of an effective cover letter add specific evidence to support the more general claims of the opening paragraph. In the case of Eva's letter, it's her chance to bring her two most important credentials, the academic and work experience mentioned in her opening, into sharp focus. (You can think of the opening paragraph as a thesis statement, and the following sections as the "argument" using specific evidence to support the thesis.)

You definitely want the potential employer to know right away that you are a university student. And if you have a major field of study and some work experience relevant to the position you are seeking, this is absolutely the most important information to convey.

If you are a liberal-arts major, the early sections of your letter might mention your research and writing skills, your practice in defining and solving complex problems, your experience in interpreting complex information and ideas, your experience with team projects, and so on. Be sure to provide specific evidence for the skills you choose to highlight: for example, your success with a major research project, with a team presentation, or with a case study or problem set. Be sure to weave in evidence of your analytical, interpretive, and writing abilities. Then consider whether your internship or job experience demonstrates how you applied some of these skills effectively. While career credentials may seem more obvious for students with a business, economics, or computer-science major, liberal-arts students can also make a strong case for their critical-thinking capacities, their understanding of different human situations, and their readiness to research solutions to problems relevant to various work situations. (For sample cover letters and résumés from a liberal-arts student and a business/economics student, see pages 69–74.)

To continue with Eva's revised cover letter to Aces Accounting, notice how the writer presents telling details about both her academic and job experience. She provides specific evidence of her broader claims about her background:

In the summer of 2014, I worked as a bookkeeper's assistant for Genesis Financial Services, a job that introduced me to many of the areas listed in your advertisement: I helped the head accountant review the general ledgers and correct journal entries, performed rudimentary account analysis, and reviewed bank statements from several different companies. I also honed my skills for interacting in person, over the phone, and through e-mail with a variety of clients.

As my attached résumé makes clear, my major course work at Cleveland State has included several beginning and intermediate accounting courses, advanced courses in financial management, and an enriched context of microeconomics and macroeconomics courses. This formal training in financial skills and economic theory was put to the test during my summer internship and has provided me with a solid foundation for my bookkeeping work thus far and for the new challenge of managing accounts at Aces Accounting.

In contrast to the earlier version of the letter, Eva now describes the main responsibilities of her summer bookkeeping job and then lists the most relevant courses from her academic training. The key point to remember is this: always foreground the *evidence* in your cover letter and avoid unsupported claims and generalities. Try not to make vague statements such as "I am very detail oriented"; rather, provide concrete evidence of your attention to detail by offering such specifics as, "The mastery of financial details was crucial to my success as a bookkeeper at the Murray accounting firm." Don't waste time with broad claims that you possess all of the fine qualities of mind and character that every other candidate will claim to have. Instead, provide evidence that proves that you are "very conscientious," "have excellent people skills," "work effectively in a team," or "love to resolve problems in the work-place." By themselves, these claims are vague generalities with nothing to back them up. Would any applicant claim *not* to have these or similar abilities? Offer specific examples to distinguish yourself from these other applicants and to help potential employers get to know you better.

The closing. The closing of the letter should briefly remind readers of your major qualifications detailed in the preceding paragraphs, provide contact information, and gracefully thank the reviewers for their time:

I will be graduating this June with a bachelor's degree in Economics, at which time I would be available to work full-time in accounting; until then, I could work for you as many as 20 hours each week. I would bring a good knowledge base, discipline, and a strong work ethic to Aces Accounting, and I hope you will decide to interview me. It would be easiest to reach me on my cell phone (216-666-7788) or by e-mail at ewinslow@gmail.com. Thank you for considering my application.

Sincerely,

Eva Winslow

This one sentence draws the letter's strands together: "I would bring a good knowledge base, discipline, and a strong work ethic to Aces Accounting." The writer has already demonstrated through her academic and internship experience that these things are so. Notice also the strategy that might be useful in some job application situations: she offers to work part-time for Aces Accounting while she completes her degree. The employer might regard this as an enticing probationary period for a new employee, and the applicant would have an opportunity to demonstrate her best qualities as a working associate.

The applicant closes by expressing her appreciation to the reviewer(s) of her application and providing her contact information. Even though the same contact information is included in the résumé, certain redundancies in application materials are useful. If the cover letter and the résumé become separated during the review process, the hiring manager can nevertheless easily find the contact information she or he needs in either document.

Before we move on to other strategies for enhancing your job-search process, here are some reminders of key points regarding cover letters.

Cover Letter Tips

Stay focused on getting interviews. The basic goal of the cover letter is to expand on and personalize the information in your résumé. The goal of both the cover letter and the résumé is to obtain an interview in which you can speak further to your qualifications and convey aspects of your personality, energy, and commitment. You *must* make it to the interview stage, because companies will almost never hire without completing an in-person, phone, or Skype interview.

Pay attention to details. Letters and résumés must be impeccably written, proofread, and well designed. Even one typo or word-usage gaffe can send an otherwise strong application to the "no" pile. Similarly, a nonspecific or generic cover letter will fail to give reviewers a true sense of you as an applicant and is very likely to be passed over.

Get off to a strong start. The opening lines of a cover letter must contain a quick reference to one or two of your key qualifications for the job. This will encourage reviewers to move on to the following paragraphs, which provide greater detail regarding your academic and professional experiences and skills. The strongest combination of factors you can offer as a college student is your academic training plus any job or internship experience: the first suggests a firm grasp of the theoretical and conceptual, and the second demonstrates the practical experience needed to do a job. If you have not yet completed your degree, be clear about where you are in your education.

Base your letter on concrete evidence rather than on claims. For example, don't just claim to possess "savvy computer skills"; instead, explain how you enhanced and used those skills in your last job or in a college research project. Don't just claim to have a "highly organized approach to tasks"; rather, show how a recent internship or leadership position in a college club or organization honed these skills. Also, if possible tie the skills that you mention in your cover letter specifically into the duties you will be asked to perform in the position to which you are applying.

Describe some of the relevant content of your university studies to employers. Even if you did not major in business, economics, or a technology discipline, you should describe, for example, your research experience, writing proficiency, and training in problem solving.

Stress throughout the letter what you can offer the company, not what you hope to gain from being employed there. (In an internship application you can strike more of a balance between the skills you can contribute to the company and the learning experience you hope to gain.)

Close the letter with a very brief reminder of your strongest attributes. You want to leave this final impression of your strengths in the reviewers' minds. Also provide your contact information, express your enthusiastic interest in and availability for an interview, and thank the reviewers for their time and consideration. Do not introduce new information about yourself. The conclusion should be brief, simple, and courteous.

Exploring Additional Examples of Résumés and Cover Letters

Next are additional examples of résumés and cover letters, from students with varied academic backgrounds.

A Résumé and Cover Letter from a Liberal-Arts Student

In her résumé and cover letter, student Alexandra Kambur, a sociology major, highlights the breadth of her liberal-arts background in global and environmental studies, as well as her research, organizational, and volunteer experience.

ALEXANDRA KAMBUR
6022 Arroyo Avenue, Goleta, CA 93117
Phone: (415) 515-6682   E-mail: akambur@gmail.com

OBJECTIVE: To obtain the program/administrative assistant position with the Sea Change Foundation in order to gain experience working in a nonprofit organization and for meaningful environmental improvements.

EDUCATION:
University of California, Santa Barbara
Bachelor of Arts in Sociology, GPA: 3.78/4.0
International Experience: Universitat Autònoma de Barcelona, Spain, Spring 2015
Expected date of graduation: June 2016

Relevant Coursework:
- Writing Courses: Academic Writing, Writing for Global Careers, and Business/Administrative Writing
- Sociology Courses: Social Inequalities in the USA, Radical Social Change, and the Sociology of AIDS
- Environmental Courses: People, Places & the Environment; Ocean and Atmosphere; and Ocean Circulation
- Global Courses: Global Conflict, Global Religions, and Religion and Healing in Global Perspectives

Research Experience:
- I completed an independent research project during fall quarter of 2015 under the supervision of John Foran, professor of sociology at UCSB. My research was focused on increases in intravenous drug use, specifically heroin, following the collapse of the Soviet Union and Russia's current HIV/AIDS epidemic.

WORK EXPERIENCE:
Cashier, UCSB Bookstore, Santa Barbara, CA   **6/2016–present**
- Practice excellent customer-service skills
- Work collaboratively with co-workers to organize quarterly work schedules
- Process, box, and organize an average of 2,000 online textbook orders at the start of each quarter

continued

Instructor/Event Coordinator, Novato Parks,   **6/2009–present**
Recreation & Community Services, Novato, CA
- Coach gymnastics to children ages 3 through 15, class sizes of up to 15 children
- Organize weekly summer and holiday sports camps, consisting of up to 50 children per week
- Trained 5 new staff members
- Developed a comprehensive preschool gymnastics progression program
- Coordinated a health campaign, "Let's Move Novato," to promote healthy living

Teacher, UCSB Global Student Awareness Club,   **5/2014–12/2014**
Santa Barbara, CA
- Taught global and cultural lesson plans to a local third-grade class of 25 students
- Researched topics for lesson plans
- Contacted prospective schools and publicized the Global Student Awareness Club

Volunteer, UCSB Safer-Sex Peers, Santa Barbara, CA   **9/2012–6/2013**
- Confidential safer-sex peer for 60 women
- Designed safer-sex information pamphlets

COMPUTER SKILLS:
- Hardware: Macintosh, PC
- Software: Very proficient in Microsoft Word, PowerPoint, and Excel. Some experience with Publisher and Photoshop
- Social Media Networks: Very proficient in Facebook, Twitter, and Instagram

LANGUAGE SKILLS:
- Fluent in written and spoken Spanish as well as English

6022 Arroyo Avenue
Goleta, CA 93117
akambur@gmail.com

April 25, 2016

Sea Change Foundation
PO Box 2929
San Francisco, CA 94126

Dear Employment Committee:

I am writing in response to the employment opportunity posted on idealist.org. I would like to be considered for the program/administrative assistant position. I am currently a student at the University of California, Santa Barbara, and will be graduating in June 2016 with a Bachelor of Arts in Sociology. I firmly believe that our earth is in crisis, and I have made it a lifelong commitment to preserve resources and encourage others to do the same. I want to be a part of the Sea Change Foundation's work to achieve meaningful social impact. I have enclosed my résumé with this letter.

After reading the job description and qualifications, I think that I have the skills and experience necessary to contribute to the Sea Change Foundation:

Research, Group Projects, and Professional Writing. I am very well organized and detail oriented. I have experience writing professional memos; creating brochures, manuals, and PowerPoint presentations; and giving oral presentations. In addition, over the past four years at UCSB I have completed extensive sociological research and numerous group projects. When doing research I plan ahead, prioritize my time, and work independently, while group projects have taught me to work collaboratively and creatively with others.

Cashier and Customer Service. I have worked at the UCSB Bookstore for the past year. During this time I have mastered all cash-register processes and have practiced excellent customer-service skills. At work I am responsible for keeping the checkout area clean and orderly, completing all cash-register processes, and, most important, attending to all customer needs. In addition, at the beginning of each quarter I work with my co-workers to complete over 2,000 online textbook orders. This includes processing, boxing, and organizing the online orders.

Teaching, Organization, and Project Development. I have six years of experience working for the City of Novato Parks, Recreation & Community Services Center as a gymnastics instructor and an administrative camp organizer. As a gymnastics instructor, I teach children and young adults ages 3 through 15. As an administrative camp organizer, I am responsible for organizing and running weekly summer sports camps. This includes designing the camp activities schedule, collecting emergency forms and contact information, designing brochures to advertise the summer camps, and overseeing all of the camp staff. This past summer, I was also a member of the leadership committee responsible for organizing an end-of-summer campaign to support healthy living in our community. Through the "Let's Move Novato" campaign we were able to bring the community together to promote healthy eating, organic food options, and exercise.

I would love to speak with you about my qualifications for the position with the Sea Change Foundation. If you have any questions, please feel free to contact me at the mailing or e-mail address listed above.

Thank you for your time and consideration,

Alexandra Kambur

A Résumé and Cover Letter from a Business/Economics Student

In the next example, Andreas, an international student, highlights his extensive work history and the very practical skills gained from his major in business, economics, and accounting.

Andreas Nitsche

School
552 University Road
Santa Barbara, CA 93106
(805) 891-8555
admissions@sa.ucsb.edu

Home
44 Willow Lane #201
Goleta, CA 93117
(805) 233-8769
andreas_nitsche@umail.edu

EDUCATION

Bachelor of Arts in Economics and Accounting (BA expected June 2016)
University of California, Santa Barbara
Grade Point Average: 4.0 out of possible 4.0

Relevant Course Work:
Intermediate Microeconomic Theory, Intermediate Macroeconomic Theory,
Intermediate Financial Accounting, Financial Statement Analysis,
Business and Administrative Writing

Associate in Arts Degree (AA, 2014)
Santa Barbara Community College
Grade Point Average: 4.0 out of possible 4.0

Meisterbrief in Timber Processing, and Construction Techniques (2012).
Grade: 2 (German grading system 1–6, with 1 being the highest grade.)

Relevant Course Work:
Business Writing, Cost Accounting & Budgeting, Law & Ethics

Vocational Education as Timber Processing Mechanic (2010)
Grade: 1 (German grading system 1–6, with 1 being the highest grade.)

WORK EXPERIENCE

Math Department, Santa Barbara City College (Fall 2014–June 2015)
Part-time Student Assistant

- Tutored college students in algebra and trigonometry
- Graded student homework assignments for the teacher

Klenk Holz AG, Bayreuth, Germany (Fall 1999–July 2005)
Production Manager

- Supervised 10 co-workers and prepared their production work schedules
- Organized and maintained a smooth and efficient production process
- Provided overall management of Inventory Control and Documentation
- Assisted in creation of improved production processes

AWARDS

- Dean's List, Santa Barbara City College (Fall 2010–Summer 2012)
- President's Honor Roll, Santa Barbara City College (2011 and 2012)

- Invitation to join Honor Society Phi Theta Kappa at SBCC (2011)
- Invitation to join Honor Society at UCSB (2010)

SKILLS

- Fluent in German and English (speaking and writing)
- Proficient in Microsoft Excel, PowerPoint, and Word

---

44 Willow Lane #201
Goleta, CA 93117

April 18, 2016

Mr. Stefan Doehmen
Analytic Jena AG
Konrad-Zuse-Strasse 1
07745 Jena
Germany

Dear Mr. Doehmen:

I have recently learned from Ms. Ingrid Zimmermann, a former leading researcher at Analytic Jena, that you provide internships within your controlling department for junior university students. I believe my previous work experience and academic training as a business and accounting student at the University of California, Santa Barbara, make me an ideal candidate for a summer internship at Analytic Jena AG.

As you will see on my résumé, accepting challenges is the foundation of my life experience. In my work as production manager in one of Germany's biggest timber-processing companies, Klenk-Holz AG, I enhanced my interpersonal communication and leadership abilities as well as the ability to meet tight production deadlines and target numbers. For example, as production manager, I led a team of 10 colleagues for whom I prepared customized work schedules in order to improve the work climate and also worker productivity. In order to meet deadlines and target numbers, I developed new production schedules that decreased the downtime of our production line by more than 25 percent. This firsthand experience, at the production level, gave me valuable insight into the core functions of a company.

Following this vital practical experience, I furthered my academic education in Germany and obtained a "Meisterbrief," the highest vocational credentials awarded by the state. This education gave me a deeper understanding of business practices within my area of professional interest — including cost accounting, budgeting, and advertising — and built the foundation for my current academic education in the United States. Moreover,

continued

extensive course work in economics, business, and accounting at the University of California has provided me with the theoretical knowledge of my desired future profession in the field of finance.

Now I would like to test and extend my knowledge in accounting by engaging in an internship with Analytic Jena AG. I am confident that this internship would be of mutual benefit. With my previous practical experience and growing academic skills, I am sure that I will be able to assist the members of your department, and, conversely, I will gain valuable experience in a field that I find both challenging and rewarding.

I would be delighted to have the opportunity to talk to you about an internship during the summer at Analytic Jena AG. If you have questions or would like to contact me, you can reach me at 806-232-8768 or online at andreas_nitsche@umail.edu. Thank you for your time and consideration.

Sincerely,

Andreas Nitsche

Applying What You've Learned

Considering the advice and examples presented earlier in this chapter, and the suggestions provided in this section, work through the following activity.

APPLICATION 3-A

Create Your Own Cover Letter and Résumé

The background. At many points in your life you will need to make a persuasive case, in writing, for your abilities and accomplishments—and not just when you are applying for a job. For instance, you might want to be considered for a special project team in your organization. Or you might apply for a training program that will help advance your career. Alternatively, you might submit an application for a research grant or for admission to a graduate school.

In this application, you will write a résumé and cover letter for a job that will help launch your career. (The assumption is that you are a senior and will graduate with a baccalaureate degree at the end of the current academic year.)

Here are some guidelines:

- Using print, online, or career-services resources on your campus, identify a posting for a real and current job that would use and further develop your academic and other qualifications. Don't make up a position or use a job posting to which you have applied in the past.

- The posting should be for a job of substance, not for the types of work that undergraduates generally undertake to help meet their educational expenses (for example, as a waitperson, a retail clerk, a camp counselor, a receptionist, or a pool lifeguard).

- Prepare a résumé that highlights the education, abilities, and practical experience that would qualify you for this particular job.

- Tailor your cover letter to the essential qualifications listed for the position. Try also to convey your personal energy and commitment in the letter, for you don't want to offer just a dry recitation of facts. Include a copy of the job posting or description with your cover letter and résumé when you turn in the assignment.

If you are just starting your career, you may be uncertain about the practical value of your academic background and work experience. So making a strong case for yourself might seem more challenging than it will be later in your professional development.

The purpose. Your immediate goal is to create a compelling picture of your work-related and personal qualities that will grab the attention of your audience and land an interview. The cover letter must make a strong impression even in the opening paragraph; otherwise, a busy reviewer might not read any further. One basic purpose of the cover letter is to get the reviewer to take a look at the credentials detailed in your résumé.

A personal, exploratory purpose of this activity is to help you think about yourself as a person who will soon begin a career. What do you hope to learn, gain, and contribute to your field? What are your specific interests and abilities, and through what sort of work might they be exercised, strengthened, and rewarded?

The audience. Imagine one or several reviewers who have received a few dozen or even more applications for the open position. In their initial screening, the busy reviewers might spend as little as 20 to 30 seconds glancing at each letter and résumé in the stack. They will quickly discard materials that don't grab their attention in a positive way. They will expect your writing to be clear, concise, and well organized. They will also expect to see specifics about you that are supported by evidence and that connect with what's needed for the advertised position. They are also very likely to shuffle you to the "no" pile if there are any misspellings, typos, grammatical errors, or document-design flaws in your materials. As noted earlier in this chapter, the review process is quick and ruthless, and you need to offer your very best thinking and writing to survive the first cut.

The communication strategy. Clarity, coherence, brevity, and attractive document design are the keys to crafting a persuasive cover letter and résumé. Do a bit of research so that you know something about the company to which you are applying, and shape the letter and résumé accordingly. Try also to determine the name and title of the person to whom you will be writing.

Don't claim anything about yourself that is untrue or inflated, but do think about the skills, training, and experiences that would make you a strong applicant for this particular job and company. One challenge is to assess and then highlight the qualifications you actually possess, especially if you are relatively new to the world of work.

Getting from the Application to Success

Assessing your skills, crafting an effective résumé and cover letter, and applying to suitable open jobs are key parts of the job-search process, but they aren't the only ones. Next, we'll examine other key steps to achieving success, in a near-term job hunt and during the coming years of your work life.

Following Up on Your Application Materials

To compete in today's economy, you need to remain an active, visible candidate—not just sit on the sidelines and wait for your merits to be noticed in your paper or online application. Follow up with a phone call or e-mail to any potential employers approximately a week after submitting your applications. Briefly let each employer know that you remain very interested in the open position and that you are available should the reviewer have any questions.

If you have other evidence you can submit regarding your qualifications, send this along. For example, can you report findings of a major research project or announce a recent leadership accomplishment? Do you have a new letter of reference for this particular job opening? If you have already assembled a paper or online portfolio (see page 44) but have not yet submitted it, be sure to forward the materials or the Web site link to the person organizing the review process. Even consider dropping by your potential employer's office in person to say "hello." (This is a risky tactic, so be careful not to be insistent or intrusive.) Your goal in all of these follow-up strategies is to put your face and voice with the more anonymous paper materials, to keep your candidacy alive.

Acing the Job Interview

While these job-interview tips extend beyond the business-writing scope of this book, you should still be well prepared for the entire job-search process, and this includes the interview.

Do your research. Research the organization with which you are interviewing. You need to have a clear understanding of its goals, products and services, recent history, and organizational structure. Such background will allow you to ask and answer questions intelligently during the interview and to make a strong impression. You should be able to find most, if not all, of this information on the organization's Web site. If you can't, see whether an externally created analysis of the company is available online. For larger established companies, check such databases as Business Source Complete and LexisNexis. Or check social-media sites, such as Facebook and LinkedIn, for company profiles and assessments.

Dress the part. Dress appropriately for the interview—certainly no T-shirts, shorts, or flip-flops. While dress standards vary widely across companies, by functions within the same company, and even in different regions of the country, you should err on the side of professionalism and dress more formally if you're unsure. You might get some hints by viewing photos on the company's Web site or by asking the human resources department what is appropriate. While business-casual attire will work for most situations, remember that you want to make a great first impression.

Be prepared. Arrive for your interview 15 minutes early to get settled in. Bring an extra copy of all the materials you submitted with your application, including your résumé, cover letter, and, if applicable, your portfolio of previous work. Bring a pen or pencil and a small notepad to take notes on any important details you learn during the interview or to write questions as you think of them. (Typing notes into a laptop during the interview would be too distracting.) Simply demonstrating your ability to arrive on time and prepared will set the right tone for your interview and immediately demonstrate your professionalism to the interviewer.

Expect difficult questions. Some interview questions are likely to be open-ended. This allows the interviewer to evaluate your ability to think on your feet, to speak confidently, and to assess your own abilities and how they will contribute to the organization. Here are some examples of open-ended interview questions:

- What interests you about this position?
- What are your greatest strengths?

- What are your weaknesses?

- Where do you see yourself professionally in five years?

Interviewers also often ask questions that pertain specifically to your possible role in the company:

- What made you apply to our organization?

- How do you see yourself fitting in at the company?

- What will you bring to the position and to the company?

Many interview strategies have shifted in recent years away from such broad questions toward targeted "behavioral" questions, such as "What would you do if . . . ?" or "How might you respond to situation X?" Such questions can be hypothetical or draw on your previous work experience. Here are some examples:

- Give us an example of a work project that you found challenging.

- Give us an example of a time when you worked effectively as part of a team.

- What would you do if you encountered a setback and could not meet a deadline?

- Have you ever interacted with an angry customer or a difficult person? How did you handle the situation?

In advance of the interview, consider how you might respond to questions like these: What *are* your broader career goals and qualifications? What examples could you present regarding work-related situations?

Conclude on a strong note. Be prepared for the most typical concluding interview question: "So, do you have any questions for us?" Avoid simply saying, "Nope, you have covered everything." Prepare a few questions in advance that you might raise regarding this particular organization. Write them down as you research the company or as they come up during the interview. For example, ask a follow-up question about a company project your interviewer mentioned to get a better sense of some of your potential day-to-day duties: "You mentioned the ongoing refinancing project. Could you tell me more about that?" Or, "You mentioned the great cooperation among co-workers here. Could you tell me a bit more about the company culture?" Asking pertinent questions will not only better inform you about the position for which you've applied, but it will also create a favorable impression of you as an enthusiastic, curious, and engaged applicant. Don't ask about salary ranges, vacations, or other perks; rather, raise questions that show your interest in doing a great job.

Stay in touch after the interview. Be sure to write a follow-up note or e-mail to the interviewer or interview group after your meeting. Express your gratitude for their time and consideration and reiterate what you like best about the organization. This simple act reflects your professionalism and goodwill as a colleague and allows you to express your ongoing interest in the job and the company. Here is an example of a follow-up note:

Dear Ms. Garcia:

I greatly enjoyed meeting you and your colleagues in our recent interview. I am more interested than ever in working with Framish Corporation and believe I could contribute the creativity and hard work you expect from all sales associates. Please don't hesitate to ask me further questions or to request more information on my background and interests.

Sincerely,

Jason Stornk
453-777-9090
jstornk@gmail.com

Your goal with such a follow-up note is to let the company know that you appreciate the time the interviewers afforded you and that you remain interested in the open position. As you consider further contacts with the company, especially if weeks pass with no word, be careful not to be so persistent that you irritate your interviewer(s). It's a tricky balance to achieve.

Following is a quick review of ways to prepare for and engage in a successful interview.

Interview Tips

Research the company to which you are applying in sufficient depth to answer questions in an impressively informed manner. Preparatory research will also help you raise your own questions that demonstrate your understanding of the company's goals, products, and services.

Dress appropriately for the interview in order to present yourself respectfully and as a serious person who will fit into a professional environment.

Show up for the interview 15 minutes early and with a notepad ready to record your thoughts and questions as the interview progresses.

Plan ahead for both the open-ended and behavior-specific questions you are likely to field during the interview. Such planning will increase your confidence and enable you to offer focused answers to questions.

Prepare some insightful questions you might raise toward the conclusion of the process, as you are likely to be asked whether you have any further questions for the interviewers.

Mail or e-mail a thank-you note after the interview. This will demonstrate your gratitude to the interviewers and indicate your continuing interest in the job opening.

Starting and Pursuing a Rewarding Career

You don't want to end up being one of those people who "lead lives of quiet desperation" (Henry David Thoreau, *Walden*). Just how to steer one's way toward and through a rewarding career is well beyond the scope of this book. However, as you begin your professional journey, keep the following things in mind:

1. Your first job or two might not be directly in line with your talents and interests, but every job you hold has some value in its own right and can contribute to the next step in your career. At the very least you will find out more, in every job, about yourself and what truly rewards you. You might develop new skills as well.

2. Unless economic necessity forces you to obtain whatever employment is available, don't squeeze yourself into a job that you already know will be a dead end. If time and money allow, seek a position that has at least a few of the responsibilities that fit with your abilities and interests.

3. In the same vein, don't fake your way into a "good" job that you know will be unrewarding to you. If the "fit" is clearly not right between you and the position offered, or between you and the company, your submitting a persuasive cover letter and pretending your way successfully through an interview might get you hired into something quite wrong for you.

4. Always remain vigilant for new opportunities within the organization for which you work. Raise your hand when an interesting task needs to be performed and you believe you can figure out how to

do it. You can't say "yes" to every request and opportunity, but you can selectively take on a project that uses your talents, hones some of your skills, and gets you noticed.

5. Especially once you have a clearer vision of where you want to go in your career life, make a tentative plan for achieving that goal and seize appropriate opportunities when they arise. Apply for positions within the company that would move you along a rewarding career path.

6. Whatever your line of work, keep researching the type of work you are doing and the industry sector that your company occupies. Especially pay attention to the changing landscape of new technologies related to your career area and to the shifting challenges and opportunities to which your company will need to respond. In the twenty-first century, no field or profession stands still for more than a few months; thus, your remaining vigilant and adaptable is crucial to your continuing career success.

The checklist on the page that follows provides a convenient summary of the steps you need to consider for a successful job search.

CHECKLIST
Overview of the Job Search

For your convenience, here is a quick summary of the key aspects of the job-search process.

☐ Make an inventory of your skills and accomplishments that pertain to your professional aspirations.

☐ Based on the inventory, create a paper or online portfolio or Web site dedicated to your professional credentials. Consider posting a filmed "interview" with yourself on your online site.

☐ Find available job postings, whether published in print or online or discovered through your personal network.

☐ Write a clear and compelling résumé and cover letter.
- Design a résumé that is attractive, accurate, and easy to read.
- Tailor the cover letter to the job you are seeking.
- Mention your key qualifications in the opening paragraph of the letter, and develop and illustrate those qualifications in the body of this document.
- Foreground the evidence; don't rely on unsupported claims about yourself.

☐ Deliver your application materials in person when possible.

☐ Make a follow-up phone call or send an e-mail to ensure that your application was received, to ask any questions you might have, and to stay visible to the employer.

☐ Keep a careful record of all your job applications and follow-up communications.

☐ Submit supplementary materials and references if requested by the employer.

☐ Prepare for the job interview.
- Research the company.
- Prepare and submit a hard-copy or online portfolio *before* the interview.
- Expect some hypothetical questions, some based on your past experience, and some broad, open-ended questions.
- Consider your own questions to raise, especially to conclude the interview.

☐ Send a follow-up thank-you note and an expression of continuing interest in the job.

☐ Consider using other strategies to enhance your credentials and presentation:
- Seek an internship related to your area of job interest.
- Enhance your formal training by, for example, building computer skills or language fluency, studying abroad, or obtaining a certificate or a "minor" in some additional program of study.

CHAPTER 4

Business Document Design, Formats, and Conventions

Understanding Key Features of Document Design

To be an effective business writer, you need to know the basics of document design, standard document formats, and business writing conventions. The design and format features include the formal characteristics of documents—their visual style and their overall design structure—that help make them clear, easy to read, and attractive to the reader. There are also a number of writing conventions to learn, assumptions about style and format that are generally shared among business professionals.

As we discuss writing formats and conventions, keep in mind that customary business communication practices change over time, and you need to remain alert to these changes. However, the underlying goals of professional communications remain fairly steady, and we'll discuss those next.

Previewing Design Basics

As you compose and design a document for a business purpose, keep in mind that your central goals are to

- produce documents, whether short or long, that communicate clearly and concisely.

- create documents that are attractive and readable and that clearly guide the reader through the piece without distracting from its content.

- make sure that the documents, whether brief e-mails or long reports, reflect your conscientious work habits and your commitment to quality.

As you aim to meet these design goals, be aware that some companies have created their own style manuals and expect employees to use formats and conventions established by the company managers. Sometimes a company's style conforms to its "brand" or identity decisions. A company might use different font styles for different purposes or different types of stationery for different internal or external audiences, or it might designate exactly when to communicate with a paper letter or through some digital format. The company might also want to preserve a certain visual consistency throughout its communications. For example, the company might want its trademark (or perhaps the color scheme associated with the trademark) to be used in designated places in certain documents. Thus, whenever you join an organization, be sure to find out if there are published guidelines for the writing tasks you might encounter. (See, for example, the *Berkeley Editorial Style Guide* or the *Wikipedia: Manual of Style*, both available on the Web.)

This chapter of *Business Writing Scenarios: Writing from the Inside* examines only the essentials of document design and business writing formats and conventions. For a more detailed treatment of the different genres of business documents (and of the important elements of grammar, word choices, sentence structures, and documentation), consult a good handbook such as Alred, Brusaw, and Oliu's *The Business Writer's Companion* (Bedford/St. Martin's, seventh edition, 2014). The Web also offers some useful resources for checking business formats and conventions:

- The "Workplace Writers" section of the Purdue Online Writing Lab is one of the best online resources available.

- The Business Writer's Free Library offers guidance on a wide range of business writing genres.

Melding Structure and Purpose

As a business writer you want always to create documents that are easy to read and understand and that are aesthetically pleasing (or at least not off-putting because of sloppy or inconsistent design).

Elements of Effective Design

The pieces that you create, whether print or digital, should always reflect the care and thought you took to compose and format the message. Document design in this broad sense includes

- the font (Times New Roman? Arial?) and font sizes (12 point? 16 point?) that you choose in order to provide clarity and emphasis.

- the emphasis that you choose to give to different words, headers, and sections of the document, decisions that include your use of italics, underlining, and boldface type. These type styles must be used consistently throughout your document.

- visual and informational cues, such as titles, subheadings, and bullet points, that guide readers through the document and emphasize key points.

- the clear labeling, numbering, and captioning of any graphs, charts, or images.

- in longer documents, the clarity and visual design of title pages, the clear titling of the main sections or "chapters," logical paragraph and page breaks, and the numbering of pages.

Through all of these complexities of form and format, your goal is to

- make your document aesthetically pleasing.

- help your reader know what she or he is reading or viewing and why.

- aid the continuity or "flow" of the document as a whole.

In longer documents, such as business plans or financial projections, you want to help readers move easily from one major section to the next and to enable busy professionals to skim or even skip sections that are not as relevant or interesting to them.

An Example of a Clearly Designed Memo

Consider the design and format choices evident in the following relatively brief document, by a student writing from the perspective of a chief executive officer. This student, Courtney Steele, is conveying very sensitive information concerning a reduction in the company's support for employee health-plan costs. Not only does she want the tone of the memo to be just right (respectful and sympathetic, reassuring when possible) and the evidence for this financial decision to be clear and persuasive, but she wants the overall design of the memo to be clear as well.

MEMORANDUM

TO: ABC Employees
FROM: Courtney Steele, Chief Executive Officer
DATE: June 1, 2016
Subject: Medical Benefits Plan Costs

Rising Costs of Health-Care Plans

With a sluggish economy and rising health-care costs, many successful companies have made the tough decision to cut or reduce employee benefits. ABC Incorporated, however, has consistently maintained our company's philosophy throughout the economic slowdown and continues to support the well-being of all employees. Unfortunately, extensive financial analysis indicates that our plan has become significantly more expensive during the past several years, and ABC Incorporated cannot continue to pay *all* of the costs of the medical benefits, as we have for the past two decades. In order to preserve our exemplary medical benefits plan, however, a new financial strategy will be implemented. Most important, our company will absorb most of the rising costs and will continue to pay the majority of the medical-plan expenses.

The New Financial Strategy

Expert financial analysts have conducted research over the past three years and have found the following data pertaining to our medical benefits plan:
- Annual costs of the benefits plan have increased, on average, about 47 percent (from an average of $8,500 per employee to a new average annual cost of $12,500 per employee).
- Total annual costs of the plan have increased by $1,200,000 over the last three years.

ABC Incorporated has clearly experienced major increases in costs. Despite these increases, our company can continue to cover 80 percent of these costs without cutting or reducing benefits. As stated before, the new average cost of our medical benefits plan is now $12,500 per employee. Additional costs not covered by the company will be distributed based on the following scale:
- Individual employees will pay, on average, $2,000 per year for full medical coverage.
- Employees with a spouse/domestic partner OR dependent child (family of two) will pay, on average, $3,000 per year for full medical coverage.
- Employees with a spouse/domestic partner and dependent child, OR two or more dependent children (family of three or more) will pay, on average, $4,000 per year for full medical coverage.
- As with the previous plan, all employees must make a co-payment per office visit. The cost will now be $15 per visit (increased from $10/visit).

Annual costs are also allocated based on income level in proportion to annual salary. The effect of costs as a proportion of salary equates to higher costs to employees with higher salaries, and consequently lower costs to employees with lower salaries. A document has been attached with a detailed chart showing exact costs for each salary level and family size. The new financial strategy will be in effect on August 1, 2016.

Questions and Concerns

Our company wants to ensure that all employees have the ability to adapt to this new financial strategy. The Human Resources Department has prepared an FAQ sheet and a prospective payment schedule plan (attached to this document) but will also be available to create a personalized payment plan that is suitable to your financial needs, or to address any other questions and concerns. Please feel free to contact our HR Director, Tom Roberts, via e-mail (HRdirector@ABCinc.com) or by telephone (extension 123). In addition, an information session will be held in the meeting room at 6 p.m. on Tuesday, June 14, where benefits experts and financial analysts will be able to answer additional questions.

Although the new financial strategy is not ideal given its financial costs to employees, we found it absolutely essential to maintain our benefits plan and continue to place the highest value on the well-being of all employees. In addition to providing an exemplary medical benefits plan, we want to be sure that all employees can acclimate to the new cost structure. We hope the attached documents and the information session will ease your adjustment to the new financial strategy and, again, please do not hesitate to contact Tom Roberts in HR with any questions or concerns.

Why the Design Succeeds

Along with her control of tone, lucid writing, and marshaling of strategic evidence in this negative-news memo, Courtney Steele has made careful decisions about the document design. The address information at the top of the memo is clear and aligned, and the subject line makes clear the content of this important communication. She then uses three subheadings to clarify the contents of the piece and bullet points to highlight critical information. The subheadings indicate the background for the changes being announced ("Rising Costs of Health-Care Plans"), the financial impact on employee medical plans and reassurance concerning cost sharing ("The New Financial Strategy"), and a plan for guiding employees' benefits choices ("Questions and Concerns").

While nothing can soften the impact of the message itself, the writer's design and format choices make it easier for the readers to see exactly what is about to happen and why. There is a clear flow of logic and evidence from beginning to end and also a visual flow that aids understanding. The broader implication is that the writer is a clear-minded and sympathetic leader and has a confident grasp of the evidence and strategies presented in the memo.

The previous example is a short document, but it suggests the principles that apply as well to lengthy reports and proposals. (For more details on designing longer documents, see page 96.) As a business writer, you should in all cases guide the reader through the information and ideas of your document. Lapses in writing clarity and coherence and careless, inconsistent document design will draw attention away from the points you want to convey. And regardless of whether it is a fair assessment, business colleagues won't think well of your competence.

Exploring Common Formats for Business Documents

Next, we'll take a closer look at the formats not only of memos but also of two other common business documents: letters and e-mails.

The Business Letter

A great deal of business correspondence still takes the form of paper or digital letters. You might, for example, need to design and compose

- a formal invitation to a dignitary.
- a cover letter for a personal job application.
- a cover letter for a company grant proposal.
- an explanatory letter to shareholders.
- a rejection letter to a job applicant.
- a billing letter to an important customer.
- a solicitation for charitable contributions.
- a request for a bid on services needed by a company.
- a letter of congratulation on a colleague's promotion.
- a formal renewal of an employee's contract.

While there are several standard forms for business letters, the most common is the block letter format:

Landscapers' Supply House
2216 East Third Street   **[The sender's address]**
Mount Shasta, CA 96067

January 25, 2016   **[The date — important on business documents]**

Ms. Betty Jameson, Manager   **[Recipient's address and title or office]**
The Timeline Hotel
555 Main Street
Sacramento, CA 94203

Dear Ms. Jameson: **[Formal salutation, last name only, followed**
                                     by a colon. Use the appropriate personal or
                                     professional title for the recipient: Ms., Mr.,
                                     Senator, Dr., Professor, etc.]

It has come to my attention that invoices for The Timeline Hotel are now 90 days overdue.
Our records show that you have received two previous notices. You are one of our most
valued customers, and we would be glad to facilitate a payment schedule with you.

I hope that everything is going well at The Timeline Hotel. I will hope to hear from
you this week regarding your outstanding balance of $4,338.00. Please contact me at
(530) 666-2319 or by e-mail at slittle@landscapers.com.

Sincerely,

[signature]

Sam Little
Accounts Receivable

Block letter format

Notice that the entire letter aligns along the left margin and that the
letter includes the address of the sender and the business address and
title of the recipient. A colon is used after the salutation (Dear Ms. Jameson:), and the paragraphs are crisp and relatively short. In general, the
block letter format may be used for shorter communications.

Standard E-mail Format

The business e-mail format can be used for both formal and informal
communications. Realizing that time-constrained clients, customers,
and colleagues won't have the patience to read large blocks of text on-
screen, savvy businesspeople tend to keep their e-mails fairly short, often
no more than 200 to 300 words or one screen in length. When you are
writing business e-mails yourself, use these strategies to make the most
of the format:

- Include a clear subject line so that the recipient will immediately
  see the topic or purpose of your communication. If the message is
  particularly important, you can indicate this in the subject line as
  well, but be discriminating. You don't want to inflate the importance of your communication. Not everything can be "urgent" or
  "top priority."

- If you need to convey more complex information and perspectives,
  use an e-mail to introduce the topic, and then add the more formal

or more complicated business document as an attachment. This practice generally makes for easier reading, since you can format the attached text with your document design skills (whereas in the e-mail itself, formatting such as bold or italics is often changed or lost when the message is transmitted).

- Try to summarize the main points of your attachment in a bulleted list to make it easy for the recipient to identify the most important information.

Following is the standard e-mail format:

TO: "Marvel, James"<jmarvel@readyserve.com>
CC: jsuarez@timeline.com   **[Copy to office or person as appropriate]**
Subject: Request for landscape bid   **[A clear and explicit subject line]**

Dear Mr. Marvel: **[Salutation plus a colon]**

As we discussed over the phone last week, The Timeline Hotel is interested in receiving a bid for landscaping services from your company, ReadyServe. To provide information for your bid, we want to arrange for you to visit The Timeline Hotel in order to inspect our grounds. Let's coordinate that visit with the availability of Janice Suarez, our maintenance supervisor.

Please let me know about your availability either by e-mail or by calling me at (530) 666-2320.

Thank you,   **[A complimentary closing. It is not**
     necessary, but it will seem friendlier to
Liz Jenkins, Director of Hotel Facilities   **most recipients.]**
The Timeline Hotel

e-mail: ljenkins@timeline.com

Standard e-mail format

In the standard e-mail format, the writer has provided a clear subject line, used a professional yet friendly tone, kept the message brief, and provided the necessary contact information. The shortness of the paragraphs in the e-mail to Mr. Marvel augments the efficiency of the communication. The document design helps convey the request quickly but still in a friendly way.

If an e-mail is being sent to a group of people rather than to an individual, sometimes the e-mail will not include a salutation (Dear Mr. Marvel) but simply begin addressing the audience already identified in the "To" line. By the same token, some e-mail addressed to large groups does not include a complimentary closing (Thank you) or repeat the name of the sender. The choice is yours, but in general it sounds more respectful and more collegial to use a salutation and complimentary closing.

A careless design, even in such a brief communication, draws attention to itself and away from the message, as in the following example:

TO: "Marvel, James"<jmarvel@readyserve.com>
CC: jsuarez@timeline.com
Subject: Touching base

Dear Mr. Marvel:

As we discussed over the phone last week, The Timeline Hotel is interested in receiving a bid for landscaping services from your company, ReadyServe. To provide information for your bid, we want to arrange for you to visit The Timeline Hotel in order to inspect our grounds. Let's coordinate that visit with the availability of Janice Suarez, our maintenance supervisor. Please let me know about your availability either by e-mail or by calling me at (530) 666-2320.

Thank you,

Liz Jenkins, Director of Hotel Facilities
The Timeline Hotel

e-mail: ljenkins@timeline.com

The content of the message is the same as in the previous example, but now the subject line is less explicit, and the paragraph breaks that made the first version easier to read are missing. Even very simple design considerations like these make a difference to the reader's understanding.

Common Memo Format

The common memo format is standard for many printed memos circulated to the employees of a company or sent to external constituencies. Its design is similar to the e-mail format, but the memo format

- is usually longer (perhaps one to two pages) and more complex in content than an e-mail.

- can incorporate more special design features (which are unavailable in many e-mail systems or become lost during Internet transmission).

You need to decide whether an e-mail or print-copy distribution will have a stronger impact on readers. Is your audience more likely to read an e-mail or a printed memo? Are your readers more likely to retain one format or medium or the other for reference? The answer to such questions depends on a lot of variables—for example, your recipient's level of comfort with computer technologies and the importance of the memo's content.

The business manager in the following example decided to summarize important employee pension changes through a printed memo:

DATE: October 17, 2016
TO: All Employees
FROM: Roselia Park, Business Manager
        Centurion Production Company
Subject: **Revised Pension Benefits Eligibility and Contributions**

We are very pleased that our several discussions and negotiations over the past three months have brought new clarity and equity to the structure of pension benefits for Centurion employees. Below is a very brief summary of the agreement we reached. The following revisions will go into effect January 1, 2017, and will be retroactive to September 1, 2016.

EMPLOYEE ELIGIBILITY

After one year of full-time employment at Centurion Production Company, all employees will be eligible for a company contribution to their pension benefits according to the following timetable:

- *1–3 years of full-time employment:* Starting in the second year of employment, Centurion Production will contribute an additional 5 percent of the employee's gross salary to the MesaRock mutual funds designated by the employee for his or her pension. (Employees become eligible for this benefit after they complete one full-time year with Centurion.)

- *4–10 years of full-time employment:* At the start of the employee's fourth year, Centurion Production will contribute an additional 7 percent of the employee's gross salary to the MesaRock mutual funds designated by the employee for his or her pension.

- *11 or more years of full-time employment:* At the start of the employee's eleventh year, Centurion Production will contribute an additional 10 percent of the employee's gross salary to the MesaRock mutual funds designated by the employee for his or her pension.

We believe the new guidelines clarify some earlier confusion and provide improved support to all employee colleagues as they contribute their talents to Centurion over the years. All the details of the eligibility and contribution rules and of the mutual funds available through MesaRock are provided on our Human Resources Web site at www.pensionbenefits@centurion.com. If you have any further questions, please contact Jane Cary, our Benefits Coordinator, at jcary@centurion.com or at extension 6667. She will be glad to advise you on eligibility and on investment options through MesaRock.

Common memo format

The memo above contains serious, contractual information, and the writer wants to make the eligibility distinctions as clear and precise as possible in a brief communication (the layers of detail are provided on the HR Web site). The business manager has made document-design decisions to use a larger font for the subject line, bold or italicized words for the key information, and bullets to make the pension-eligibility distinctions as clear as possible. Notice, also, the parallel structure at the top of the memo: the information is aligned in columns for easy reading and a neat appearance—all part of good document design.

Not even the best document design can replace the necessity for clear writing and the logical presentation of evidence, information, and ideas, but structuring your business documents sensibly can help you get your message across more clearly and effectively.

Incorporating Visual Materials into Your Text

As part of document design, we should consider the increasing importance of visual materials (graphs, photos, tables, charts, and so on), especially in longer documents, such as business reports and proposals. In all cases, you want to be sure that the visuals you include are truly an aid to the reader's understanding.

Avoid Do-Nothing Graphics

Be sure that the graphics you include genuinely aid the reader's understanding or add a visual emphasis to the point you are making. At the same time, avoid visual elements that are merely decorative (and maybe even distracting) or try to demonstrate what is already obvious in the text. For example, if you had made clear in the text that your company's annual philanthropic contributions were fairly evenly divided among three organizations, the following pie chart would *not* be of any use to the reader:

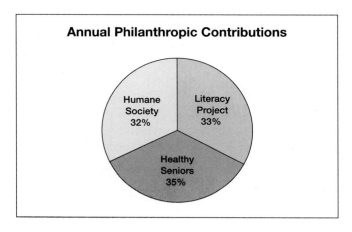

This chart tells readers nothing they can't already glean from the text and might even insult them by implying that a visual aid is necessary at all. The roughly one-third distributions are too obvious to require a visual illustration.

Also be sure to exclude purely decorative graphics that do nothing to support or illustrate your content. For example, the writer of a grant proposal could insert a flourish of the sort shown below just because he likes its art nouveau grace, but the graphic might make readers wonder about its significance, thus distracting them from the content.

Exceptions to this rule might include a company logo or perhaps a graphic that brings attention to the company's product, service, or key clients. For example, McDonald's includes its famous golden-arches trademark in many of its publications. Toyota includes photos of a few of its newest cars in its annual report just to remind consumers and shareholders of advanced design elements and new developments. Toyota also includes many photographs of its corporate leaders in an effort to humanize the company.

In general, when considering what images to include in your documents, you want to select visuals that help clarify complex data, show numerical or other changes more dramatically than would be conveyed through words alone, or illustrate a company process or structure. (The next section provides examples of such visuals.)

Provide Context

Be certain always to *contextualize* the graphics you decide to use, following these strategies:

- Introduce all graphics in your prose, letting readers know what you expect them to see in the chart or diagram that follows. (Keep in mind that many readers will simply skip over any visuals.) If the graphic falls on a page other than the one on which it's introduced, identify the appropriate page number.

- Follow the graphic with a brief reference back to a key element in it—again emphasizing what you want readers to notice. (The strategies of introducing a graphic and providing a follow-up comment on it are illustrated in the table and figure examples shown later in this section.)

- Give each graphic a clear title (and sometimes a number, as in "Figure 4-1"), and provide its source if it was copied or derived from a published source. Keep these titles and the citations consistent across all of the visuals in your document.

n the graphic if necessary. In some cases, the title will not be ..ough to explain, say, a trend illustrated in a line graph or key data points included in a table. The graphic should be able to stand alone and still present clear information to readers.

As you weave visuals into your text, remain in charge of guiding readers toward what you want them to see. Don't just let the graphic "float" somewhere in your text; rather, connect it clearly to your narrative flow. In reality, many readers will likely take your prose interpretation at face value and hardly glance at the supplementary graphic—unless you make a claim that's hard to believe or the graphic paints a clearer picture than your words do.

Following are two examples of how you might contextualize visuals in print: a table (a term referring to an arrangement of numerical data in rows and columns) and a figure (a term that covers all other graphic materials, including maps, photos, and drawings). Consider the following extract from one company's financial report.

The manufacture of these special devices in the U.S. has grown in recent years as the nation of Sukistan moved away from this market and into the production of nonrelated materials. Table 4-2 (below) shows how much the Framitz Industry has grown in annual sales from 2004 through 2013:

Year	Units Sold	Gross Revenue
2004	1,250	$185,000
2005	3,226	$260,000
2006	3,500	$433,000
2007	3,899	$490,000
2008	6,045	$899,000
2009	6,600	$1,112,000
2010	6,800	$1,133,000
2011	7,255	$1,344,000
2012	11,000	$3,455,000
2013	12,550	$4,123,000

Table 4-2: Annual Sales Growth of the Framitz Industry
(Source: Framitz Overview at www.framitz.com/322)

Notice the extraordinary revenue growth in Framitz sales in 2008 (a rise of 55 percent over the previous year) and in 2012 (a jump of 257 percent). In each of these years the United Arab Emirates purchased extraordinary numbers of Framitz technology.

In the text preceding the table, the writer makes a particular claim (in this case, that Framitz Industry's manufacturing output has increased in recent years) and then asks readers to examine this trend visually in the associated graphic. In the text that follows the table, the writer then draws readers' attention to a couple of dramatic growth figures. These specific examples serve to strengthen the writer's assertions regarding Framitz Industry's growth in revenue.

Figures (everything other than a table) must also be contextualized. In the following report extract, an organizational chart illustrates something seemingly simple (a company's organizational structure). Yet the text that precedes and follows the chart makes the message of this visual clear: this company's management structure is creating inefficiency.

The management structure of Asteroid Industries remains lean at this time, with only four major directors overseeing a workforce of nearly 2,100 employees. The organizational chart below illustrates the management structure:

Figure 4.1: Asteroid Industries organizational structure

It is worth noting that the HR manager at Asteroid Industries organizes all pension and medical benefits for the 2,100 employees, and the VP for sales needs to coordinate 235 salespeople working in seven U.S. regional offices. There is no cadre of assistant directors to support either of these two management positions.

The key to the effective use of graphics is to make sure the graphics are an integral part of your document's narrative flow. Don't allow the graphics to float on the page disconnected from the rest of your text. Provide introductory context, and then follow the graphic with a brief analysis of what the graphic illustrates.

Designing Longer Documents

Regardless of the length of your document, your design goals should be the same: to aid reader understanding. Be consistent in the decisions you make about font styles, text breaks, headings, subheadings, graphics, and other design features that allow readers to recognize at a glance where they are in the document.

In longer, more complex documents, some additional design considerations come into play. For example, business reports, which can run perhaps 15 or more pages, may include special sections that introduce and summarize the content of the piece, that outline current markets and sales strategies, or that propose new directions for the company. These reports may also include financial forecasts and concluding recommendations. To provide a coherent reading experience, a document of this sort will need to include

- a clear and explicit title.
- an indication, perhaps on the title page, of the report's author(s) and date of completion.
- a table of contents.
- a brief overview of the document's content (often an Introduction, Executive Summary, or Abstract).
- headings and subheadings to orient the reader.
- dividers between major sections of the piece. (This could mean a new heading for each new section, a new title page for each, or, in some cases, a title page plus dividing tabs.)
- a formal conclusion that pulls all of the document's sections together. (Sometimes this section will make recommendations as well.)
- a Works Cited or References page that uses a standard bibliography format (often MLA, APA, or *Chicago Manual of Style*) to acknowledge major sources of information used in the document.

For these long documents, structure and organization are especially important concerns. Creating a clear and consistent design structure is time consuming, and you should plan ahead for a significant time commitment from you, from your team, or from both. It's especially helpful to plan the design aspects of a longer report even before you write the sections. At the very least, you should decide on font sizes and styles for the various sections and subsections you plan to write. For examples of large-scale document designs, see Chapter 7, "More Complex Business Writing Projects."

Following Other Style and Format Conventions

In addition to the elements of document design presented earlier in this chapter, there are other important conventions that characterize professional discourse. Conventions are common writing practices that have developed over the years among business professionals. While conventions often make their own sense with respect to clear communication, they also constitute a set of shared expectations within and among professional communities. They are a signal system that says, in effect, "We belong to a professional community with agreed-upon communication practices."

Some of the more common conventions of business writing are discussed below. Be aware, however, that conventions can change with time, and you need to remain alert to these changes. (One convention that does not change over time, however, is the requirement that writing be clear and error-free. For advice on using clear vocabulary and on proofreading, see pages 27–29.) Further, as stated earlier, some companies have created their own style manuals and expect employees to use formats and conventions established by the company managers. The next few pages discuss some common conventions of business writing.

Salutations, or "You Lost Me at Hello"

You have probably noticed in ordinary social interactions that some people might seem too chummy or aloof in the way they greet you. For business writing, there is a well-developed protocol for greetings or salutations that show respect or friendliness, or both, toward the people you are addressing. The examples that follow demonstrate common appropriate—and inappropriate—salutations for business writing.

Appropriate Salutations

Dear Ms. Rodriguez:

The use of "Ms." followed by the person's last name and a colon is nearly always the professional manner for addressing a female recipient. The social and political arguments over the use of "Ms." were decided long ago, and "Ms." has become the professional standard.

Dear Miss Rodriguez: or Dear Mrs. Rodriguez:

Use "Miss" or "Mrs." in a salutation only if you know that the recipient prefers to be addressed with these titles. For example, if you see the recipient sign her letter as "Mrs. Ann Henderson," you know that she prefers that title. If a bank teller's name plate says "Miss Henderson" or "Mrs. Henderson," you can be sure that she has stated this preference to the bank managers. In the United States in the twenty-first century, all of us get to choose how we want to be addressed.

Dear Mr. Nelson:

"Mr." followed by the person's last name and a colon is appropriate for 90 percent of the business correspondence you will address to men. However, when your recipient, male or female, holds a particular office, professional status, or distinction, it is often appropriate to use her or his title in the salutation, followed by a colon.

Dear Senator McCain:
Dear Mayor Schneider:
Dear Professor Rodriguez:
Dear Dr. Park:

According to conventional practice, it is best to use formal titles in place of Mr. or Ms. when the recipient is a PhD or an MD, is a professor, holds a military rank, or holds a political position. Some people will be irritated if your salutation mistakenly drops the professional titles they have earned. By convention, a salutation to a company CEO, director, manager, or committee chair usually does not include these titles. However, the *address* of a formal letter should:

> Norma Kaplan, District Manager
> Allied West Association
> 6645 Tenth Avenue
> New York, NY 10008
>
> Dear Ms. Kaplan:

Dear Customers:
Dear Hiring Manager:
Dear Review Committee:
Dear Human Resources Director:

When you are writing to a large, general audience, you can use an appropriate group title (as in "Dear Customers:"). If you cannot determine through phone inquiry or Internet research the name(s) of your recipient(s), do not revert to the outmoded "To Whom It May Concern" or "Dear Sir or Madam." Instead, devise a salutation like those suggested above.

Dear Serena,

The use of a person's first name, followed by a comma, is acceptable for an informal, friendly interaction if you know the addressee on a personal basis.

Hi, Jack,

This is probably the friendliest, most relaxed mode of greeting and is appropriate *only* for a casual e-mail interaction between associates who are on very friendly terms and in a nonhierarchical relationship. The "Hi" can also be used in an e-mail to make clear at the outset that there is nothing negative or adversarial in the subsequent message.

Inappropriate Salutations

Dear Serena Rodriguez, or Dear Serena Rodriguez:

Whether using the informal comma or the more professional colon after the salutation, *never use the person's full name*. We frequently see this error in messages we receive only because of the lazy use of giant merge lists (computer-generated lists of recipients for mass e-mails or paper mailings).

Jack— or Jill—

Don't leave off the customary "Dear" in a salutation unless you are on very casual terms with the recipient. These unadorned salutations can sound peremptory and disrespectful.

Jenkins—

Also don't hail your recipient with a last name alone. Especially to older generations, the use of the last name by itself is very likely to sound aggressive, as though a negative message will quickly follow: "Jenkins—I don't know how the heck you got these crazy numbers!"

Though there are many possible salutations, keep in mind that the most common business greeting is "Dear Mr." or "Dear Ms.," followed by the last name and a colon:

Dear Ms. Liu:
Dear Mr. McFall:

And be sure to spell the recipient's name correctly. It will look inattentive and careless if you mistakenly write "Smith" instead of "Smythe" or "Elis" instead of "Elyse." You want to start on a solid footing when you greet your professional colleague or prospective employer.

Type Size and Style

In general, use 12-point type in the main text of a document, a font size easy for most readers to see. The Times New Roman font has long been the standard for business communications, but it's acceptable to use other fonts, such as Arial and Calibri, providing that they're clear, easy to read, and professional in appearance. (Check to see whether the organization for which you work has made decisions about font styles for its various communications.)

In longer pieces of writing, decide what size and style of font you will use for the main section headings and for subheadings. Be consistent in your use of **bold** type, *italics*, and <u>underlining</u> throughout the document. Also, avoid using colored type unless you wish to highlight specific or important information in the text.

Margins

In general, use standard one-inch margins at the top, bottom, and sides of the page—or at least use consistent margins throughout the text. Use margins that are justified to the left. You can also use full-justified margins (both left and right) as long as the process doesn't result in odd blank spaces or s t r e t c h e d   w o r d s.

Text Breaks

In business writing it is good practice not to create long paragraphs and long sections of uninterrupted prose (for example, a full page or more). Use paragraph breaks that are appropriate to the logic of the information you are providing, but make the paragraphs short and easily digested.

In longer documents, consider using subheadings to signal a shift to a new topic. This strategy helps orient readers and allows them to skip ahead and find a particular topic. Keep in mind, though, that lots of *very* short sections of prose are distracting and can disrupt the flow of the reading process.

"Widows" and "orphans" are inappropriate breaks in a text at the bottom (*orphan*) or top (*widow*) of a page. Plan your page breaks so that little pieces of text or headings don't linger, lonely and in the wrong place. For example, you would not want to make the reader move to the next page to read the last three words in a major section of your report. Nor would you want to introduce a bulleted list with a title at the very bottom of a page and follow with all the bullet points on the *next* page. These odd textual breaks imply lazy formatting and will irritate most readers or distract them from your content.

Bullets

The use of bullets has always been controversial in business writing. Some people think bullets make central points stand out for the reader, and others think they are a poor substitute for clear narrative prose. Most business writers, however, would agree that bullet points should

- be used smartly and sparingly.
- be reserved mainly for lists of brief key facts.
- be written in parallel structure. (In other words, use the same grammatical structure from one bullet point to the next, as we did in this example.)

Pagination

Insert page numbers in longer pieces of writing. Begin pagination with the main body of the text, not with a title page or a table of contents.

Typically, page numbers are centered at the bottom of a report, but you can decide what looks best for your purposes and for the genre of the document. Just be consistent throughout.

Headers

Headers are a useful organizational tool that allow you to divide your document into short, easily digested sections. When included at the top of a page, headers also remind readers what section they are reading in a longer report or proposal or allow them to easily locate a particular section. For example, if you were writing a psychological study, you might use headings that pertain to the different sections of the study, such as Abstract, Methods, Results, Discussion, and so on.

Responding to Real-World Writing Scenarios

Let's apply some of the document-design strategies discussed previously to the following scenario.

Analyzing a Writing Scenario: Improving a Poorly Crafted Memo

The following memo is poorly designed and formatted. In addition, it contains awkward wording and typographical errors, and it does not address a sensitive issue as thoughtfully as it could. Analyze the structure and wording of this document, considering how each could be improved. As you do so, you might find it helpful to refer to earlier sections of this chapter: "An Example of a Clearly Designed Memo" (page 85) and "Common Memo Format" (page 91).

DATE: February 26, 2016
TO: Employees of Stealth Technologies
FROM: Barbara Long, HR
Subject: Computers and the Internet

It's come to our attention that some people are making too much personal use of their cell phones and the Internet during regular office hours. We don't want to prevent you from important contacts with your spouse/partner or children during the day or to curtail all personal use of the Internet, but some supervisors and staff have reported excessive distractions by some employees. We need to refocus on our professional obligations to the important work of Stealth Technologies, so here are new guidelines: First, whenever possible employees should limit there personal phone use to the lunch and break periods.

Only emergency situations—for example a child's illness at school or a need to change transportation arrangements with a spouse—should lead to personal phone calls or text messages or tweets. If you do have an emergency situation, alert your supervisor as quickly as possible. Second the office Internet connections should not be used during work hours for online shopping or other personal explorations. Do this on your own time. Anyone who does not exercise reasonable restraint in his or her cell phone and Internet use will receive a warning from their supervisor. Subsequent misuse of office time will be subject to provisions ST431–ST433 in the Personnel Manual. We hope that these reminders don't seem to stringent to any of you and we trust that everyone will want to cooperate.

Applying What You've Learned

Using what you have learned about style, format, and writing conventions in this chapter, work through the following activity, which asks you to apply your knowledge of proper design, formatting, and conventions for business documents in order to revise a poorly crafted piece of business writing.

APPLICATION 4-A

Revise a Poorly Crafted Memo

Revise the previous memo based on the guidelines in this chapter and on the following suggestions:

- Can you correct the inconsistent formatting that introduces the memo, provide a full title for Ms. Long, and devise a more explicit subject line?

- Can you make the tone of the memo more collegial, more likely to obtain the cooperation of valued colleagues?

- Can you improve the awkward phrasing in several parts of the memo and correct any typos and grammatical and punctuation errors? (For more on proofreading, see page 29.)

- Can you break up this single block of text by using subheadings or bulleted entries, or both, that will make the memo easier to follow?

CHAPTER 5

Writing to Colleagues within the Organization

Understanding the Challenges of Writing to Colleagues

The goals of clarity, accuracy, tact, and respect should shape all business writing, whether the audiences are internal or external to the organization. When communicating with colleagues in particular, it is important to address them, whenever possible, as valued partners in a shared enterprise. This style of communication will help foster a culture of cooperation, productivity, and respect that is crucial to any organization's success.

Keeping Special Issues and Controversies in Mind

When communicating with colleagues, be mindful of workplace cultural and political issues that can affect how your message is received:

- Colleagues generally have close knowledge of the company's practices, successes and failures, good and bad personnel histories, resource limitations, and so on. Though individual employees may have an incorrect understanding of these things, they may be wedded to their own assessments. For example, some colleagues may be unhappy about recent changes to medical benefits, while others might be pleased with the newer options; or the technical-support staff may feel good about several years of increased budgets, while the shipping department may have watched its resources decline.

- Staff with relatively long histories in the organization may feel either well rewarded by the company or underappreciated. For example, Joe might feel passed over for promotion, while Mary may have been recognized repeatedly for merit increases and title changes. In addition, some "political" factions in the organization may persistently press for certain changes and outcomes or may resist new initiatives. For instance, a high proportion of employees might support a new policy allowing flexible scheduling, while a small faction might complain that this change will result in some workers "picking up the slack" for others. Therefore, communicating about this policy would require special sensitivity and care.

- A shared vocabulary of buzzwords and hot-button issues might evoke a strong response among some colleagues, a reaction of which the business writer must be aware. For example, many employees might support their company's recycling efforts; others, however, may think that this initiative has been a waste of time and cringe when language associated with environmental interests appears in new company policies.

Respecting Co-workers across Business Cultures

Some businesses are more collegial and egalitarian, or hierarchical and formal, than others. Some professional areas are known for being more or less abrupt, gracious, aggressive, or accommodating in their communication styles. A business writer needs to know the culture within which she or he is operating in order to communicate effectively.

Some Good News about Workplace Dynamics

Fortunately, most businesses and professions that you, as college and university graduates, are likely to enter operate with a relatively collegial style. In these settings, many things are accomplished less by peremptory decrees than through reasoned discussion, the examination of evidence and alternatives, and friendly persuasion. A few CEOs do gain notoriety (in television appearances or in the financial press) for their aggressive, in-your-face style, but most executives would much rather have their companies appear in one of the annual lists of "the best places to work."

These leaders want to cultivate the loyalty, creativity, and productivity of their talented co-workers as they contribute to a common enterprise.

Whatever their personal value systems, most top managers consider the pragmatics of guiding an organization over an extended period of time. A CEO who regularly issues commands, upbraids or micromanages employees, and suddenly reverses direction will not be tolerated for long by talented people. Those people may hunker down for a while during a poor job market, but they will seek greener pastures when employment prospects improve. Further, effective managers understand that a business environment that treats colleagues as valued partners in a shared endeavor is much more likely to earn employees' commitment to quality and productivity. This book addresses all of its suggestions and examples to the more cooperative and more respectful mode of business leadership.

Potential Consequences of Angry Communications

Consider what happened when a highly successful and experienced CEO e-mailed an angry memo to his managers, as reported on April 6, 2001, in the *Daily Telegraph* of London. The intended audience was internal to the company, and even that audience alone would probably not have been responsive to their leader's demands and threats, shown in the following excerpt:

> We are getting less than 40 hours of work from a large number of our K.C.-based EMPLOYEES. The parking lot is sparsely used at 8 a.m.; likewise at 5 p.m. As managers—you either do not know what your EMPLOYEES are doing or you do not CARE. . . . In either case, you have a problem and you will fix it or I will replace you.
>
> NEVER in my career have I allowed a team which worked for me to think they had a 40-hour job. I have allowed YOU to create a culture which is permitting this. NO LONGER.

The CEO concluded the memo by threatening that "hell [would] freeze over" before he would increase employee benefits. He also said that he wanted to see the parking lot nearly full by 7:30 a.m. on weekdays and half full on weekends. Finally, he warned, "You have two weeks. Tick, tock."

Notice the CEO's implied shouting in capital letters and the suggestion that "EMPLOYEES" are entirely subordinate to the managers and should be severely reprimanded. The content of the e-mail was so juicy that it was quickly leaked to Yahoo! and to the financial press. In a very short time, the corporation's stock value (and thus its net worth) fell by

22 percent. Investors inferred from the e-mail that there were serious productivity problems in the corporation and that the CEO was coming unglued.

The bottom line is that this angry communication had a negative impact not only on the company's managers but also on its reputation and prosperity. Chapter 8, "Business Writing Gaffes in the Real World," addresses, in greater detail, the need to think about unintended audiences—and consequences—for business communications. This awareness is especially important today, as the boundary between internal and external, private and public, communications continues to fade.

Responding to Discourteous Communications

Let's say that you are a purchasing assistant for a supplier of computer software and hardware. Several staff members and you have spent an entire month researching the least costly means of obtaining a large number of circuit boards for resale to a big distributor. You don't want to irritate Adam Bekwik, the CEO, but you will need time to do further research on a better alternative for buying more circuit boards. You just can't have the new assessments on his desk by tomorrow, as he demands in the following communication:

Kimball—I can't figure out how you got these weird figures for our big purchase of circuit boards from J-Peg Enterprises. Have some new numbers on my desk tomorrow morning. We just can't afford such a big outlay at this time, and I trust you and the purchasing group will be able to find a less costly means of getting the items we need for resale.

Adam Bekwik
CEO

How would you respond, as Kimball, respectfully, tactfully, and with a note of reassurance? The purchasing team has already done a good deal of research, and you probably need Adam Bekwik to be more explicit about his dissatisfaction. Here is one possible response:

Dear Adam:

I am sorry that the figures we provided are not as convincing as they might be. The team will be very happy to take another close look at the numbers in order to bring the recommendations in line with available company budgets. It would help us greatly if you could provide a few more specifics about the costs you believe we could meet at this time. While we would not, unfortunately, be able to provide improved research to you by tomorrow, with your further guidance we could submit by this Wednesday a more focused report that meets your expectations. We appreciate your patience as we craft a more acceptable set of recommendations.

Sincerely,

Jacqueline Kimball, Purchasing Assistant

Kimball has responded tactfully and respectfully, not defensively, to the boss. She asks Bekwik for further guidance for the team as it pursues more research, which will probably persuade Bekwik to give the team more time. You can bet that this calm and confident response will earn respect from her boss.

Distinguishing Friends from Professional Colleagues

It's very common for some number of co-workers to become good friends, and we certainly live in times when the concept of one's "friends" has been greatly expanded through social media like Facebook. Younger workers are also used to much more casual, less hierarchical relationships with their teachers, co-workers, and bosses.

This relaxation of formalities can easily lead, however, to business communications that are too informal and unclear. Consider always that different co-workers and managers in a professional organization will have different communication expectations, and many will not be impressed by professional writing that is presumptuously familiar and includes slang, chattiness, and the vagueness that often characterize the interactions among friends when they "hang out" together or communicate over cell phones or through texting and tweeting. So be prepared to use a more formal style when you are writing in the context of your work responsibilities.

Consider, for example, the dramatic differences between the following two responses to a colleague, Jack, who has requested financial information:

(a)

> Yo, Jack—
>
> That thing we talked about yesterday won't actually be ready until sometime later. It'll be a top priority for sure, but we need to get some other stuff together before we can produce it. Stay tuned.
>
> ~ *Jill*

(b)

> Dear Jack—As it turns out, the further analysis of our 1999 tax filings that we discussed yesterday will take a few days longer than I had thought. The records for 2000 and earlier are in our warehouse and will need to be retrieved. That will take a few days, but I will make every effort to have the analysis in your hands no later than this coming Friday, September 30. Thanks for your patience on this important matter.
>
> ~ *Jill*

The tone and clarity of content are entirely different in these two responses to Jack. Response (a) is too casual ("Yo") and imprecise to get a respectful hearing. Vague references to "stuff" and "that thing" might not even help Jack recall what he asked for in the first place, and he won't be assured that Jill has any specific knowledge to share. The imprecision of (a) might be fine during a chat with a pal, but the vague references won't impress colleagues that Jill is professionally competent.

In contrast, response (b) is still relaxed and friendly, but it has a more professional tone and offers specific information on just what Jack needs to know, provides persuasive reasons for a further delay, and assures him that he will receive the needed information by a certain date. It is this level of competence and focus that you will want to convey in all of your work-related thinking, writing, and behavior.

Conveying Negative News

Disseminating bad news to one's colleagues, whether individually or collectively, is one of the trickiest and most taxing assignments you will undertake as a professional. The bearer of bad news will seldom be appreciated for her or his efforts, even if the writer had nothing to do with the decision making that led to the negative developments. It is best for your personal ambitions and for the long-term reputation of your organization to face these communication tasks, whenever possible, with honesty and with a measure of respect and empathy for your audience.

Negative News Pitfalls

Consider first some cautions on delivering negative news, all of which were derived from the experiences of professional colleagues.

1. **Wanting your audience to like you.** Just accept that very few people will like the bearer of bad news; your desire to be liked by others will cloud your communication effort. For example, your announcement regarding reduced medical benefits will get muddled if you try both to be clear in your message and to evoke your friendship connections with fellow employees. Instead, just present the bad news with a sympathetic understanding of its impact on your colleagues: "This necessary change to benefits will be a challenge to all of us."

2. **Being more dramatic than empathetic.** Certainly you should convey empathy, but don't wring your hands in shared agony. This caution is slightly different from pitfall 1. For example, recipients of your bad news on a salary freeze are likely to take their cues from you: they will probably react negatively if you present the freeze as an "awful problem for all of us" but more positively if you describe it as "an unfortunate and, we trust, a temporary necessity."

3. **Minimizing the impact or problem.** Sugarcoating, or writing euphemistically, will seem uncaring, dismissive, or evasive. For example, it will seem as though you just "don't get it" if you describe mandatory furlough days, with reduced salaries, as "a great way for you to have more time for yourself and with your family." Tell it straight but with sympathetic understanding.

4. **Obfuscating.** This tactic confuses readers and might make them wonder what you are hiding. For example, a report that attributes a sharp drop in a retailer's stock price to "usual market cycles" is unlikely to be persuasive. If, in fact, the stock value plunged because a harsh winter kept many shoppers at home, thus reducing profits, it is best to state this fact.

5. **Seeming aloof, or "Better you than me."** In this type of communication, the writer seems disengaged from the problems of others. For example, if you announce a more restrictive policy on employee parking, avoid mentioning your own designated parking space and express genuine concern for the situation of others.

6. **Burning bridges.** Whenever possible, keep relationships intact and lines of communication open. Write with as much respect and restraint as possible, even in a contentious situation. Try to stay above the fray, for example, when you receive an angry e-mail from an associate who hates your decision on the assignment of new office spaces.

7. **Trying to sweeten bitter news.** Trying too hard to be cheerful will irritate the recipients of negative news. For example, in a letter announcing more restrictive policies on vacation times, don't conclude with "Have a nice day." Instead, end with something more like this: "We will appreciate your adjustment to this new policy. Don't hesitate to talk with me if you have particular problems regarding pending vacation plans or related concerns."

8. **Clouding your purpose.** Be clear about your purposes in communicating. Are you announcing a decision or seeking input from colleagues in advance of a decision? For example, in Application 5-E (page 122), it is imperative for employees to know that you are conducting a feasibility study on the possibility of switching to flexible work hours, not announcing that you will definitely implement an opportunity of this sort.

9. **Offering too much information.** This approach risks confusing the audience and provides more ammunition for skeptics in their complaints and rebuttals. Say, for example, that a memo explaining pension changes presents a barrage of numbers, burying what employees will see as the most important news: that the company's contribution to pensions is being reduced. A better strategy would be to offer a couple of typical examples of the financial impact on employees in certain salary ranges.

10. **Providing too little information.** This approach will fail to persuade the audience and may fuel skepticism. For example, if you are announcing a reduction in the tech-support staff whom many employees rely on, you need to treat your colleagues as adults who are amenable to reason and evidence. Give them the key pieces of information that led to this decision and explain how sufficient tech support will be provided in the future.

11. **Assuming that one communication is sufficient.** In a well-run company, announcements of serious changes in policies and resources are usually preceded by planning discussions and followed by further question-and-answer (Q&A) opportunities for employees. For example, if your company is considering a cutback in pension benefits for newer employees, you will have a

better chance of not alienating colleagues if you plan a series of preliminary explorations with key staff, hold some open Q&A sessions, announce the decision with clarity and tact, and follow up with further Q&A sessions that also offer expert advice on various pension-savings alternatives.

12. **Setting (possibly) unrealistic expectations.** Making direct or implicit promises that you might not be able to keep, or hinting at better times to come just kicks the can down the road. For example, in the scenario on pages 114–17, announcing reductions in a company's financial support of employee medical benefits, the reductions are very likely to be permanent, and any hints of a "brighter future" would unfairly raise workers' expectations and set them up for disappointment.

Responding to Real-World Writing Scenarios

Let's see how some of the negative-news cautions from the previous section might be applied to the following scenarios.

Analyzing a Writing Scenario: Conveying Bad News about a Holiday Gift

Sarah Boss, the CEO at Blessmark Industries, is announcing that she can no longer provide a Thanksgiving turkey to each employee. This is not a dire situation for anyone, but her employees will be disappointed by the loss of this annual Thanksgiving gift. As many managers have come to know, it is common for the recipients of a special perk to eventually regard the benefit as an entitlement, not as an occasional generosity. I call this phenomenon *"Where's my darn turkey!"*

An Off-Putting Message about the Gift

In this first draft, Sarah Boss does a poor job of conveying her negative news:

November 16, 2016

Dear Employees of Blessmark Industries:

I am sorry to report that I can no longer afford to purchase a turkey for every employee's family. I know you have enjoyed this tradition since 2011, but times have changed. I do wish you all the best in your Thanksgiving celebrations.

Sincerely,

Sarah Boss

The writer makes no attempt to explain what has changed to prevent her from supplying the annual turkeys, and then she abruptly signs off with the non sequitur "I do wish you all the best." At the very least, her colleagues will feel they are being treated as peons whose holiday benefit has been arbitrarily suspended.

A More Thoughtful Message

Here is a second draft that is much less likely to alienate Sarah Boss's employees:

Dear Employees of Blessmark:

Many of you will recall that, in 2011, I supplied each of my 250 staff colleagues with a turkey for Thanksgiving. The purchase was made with my own funds and as a gesture of thanks to the many dedicated co-workers who had led us to prosperous times after several years of severe financial constraints.

Your pleasure in receiving the unexpected gift inspired me to provide the same token of appreciation for Thanksgiving 2012, 2013, 2014, and 2015. As much as I would like to continue this tradition, I find that I no longer have the personal or financial resources to visit the 12 supermarkets in our area that have in the past supplied me with the turkeys needed. I hope you will understand that I continue to appreciate your wonderful dedication to our company.

Please accept my thanks during this season of celebrations as we contemplate our good fortune on so many levels. I hope the happiness you experience this Thanksgiving with friends and family will not be diminished by my inability to contribute to the feast.

Sincerely,

Sarah Boss, CEO

In the revised version, the CEO writes with appreciation and sympathy for her colleagues, and she explains why she can no longer supply the turkeys. Many employees probably did not even know about the labor and personal expense involved in the effort. When the CEO expresses her thanks at the end of the letter and wishes her colleagues happiness during the holidays, she is more likely to be taken as sincere.

Analyzing a Writing Scenario: Conveying Bad News about Medical Benefits

This scenario describes a serious situation facing businesses across the United States. The company in question has for many years supplied free medical benefits to all employees but now must require employees to share in the rapidly escalating costs. This policy change will have significant, long-term consequences for all of the company's workers.

The scenario will ask you to imagine that you must deliver the news of the change to your colleagues. It will also provide tips on how to break the news effectively and accurately, with the least amount of fallout.

The Background

You are the CEO of a medium-sized company of 300 employees, and you need to write a memo informing all employees that they must start, as of August 1, 2016, to pay part of the costs for the medical benefits offered through the company. How will you convey this difficult news in a friendly and understanding manner? What follow-up strategies for employee discussion and comment will you devise as this plan moves toward implementation? How many details will you include in this memo, and how many will you choose to present through a different venue (such as Q&A meetings)?

The Basic Facts about the Benefits Changes

For 20 years, the company has been able to pay *all* of the costs of the medical-benefits plan (except for a modest co-payment), which covered the individual employee and his or her spouse and dependent children. Financial analyses over the past three years have shown that the annual costs to the company for this generous benefit have grown, on average, from $8,500 per employee plan to $12,500 per employee plan—an increase of 47 percent, or an increased annual company cost of $1.2 million, since a mere three years ago. The company cannot absorb the full cost of all of these increases.

The new financial strategy is for the company to continue to absorb some of the current and future increases but also to pass some costs on to employees. In addition, the new system will charge less for individual employee coverage and more for spouse and dependent coverage. Here are the details:

- Individual employees will pay, on average, $2,000 per year.

- Employees with a spouse OR dependent child (family of two) will pay, on average, $3,000 per year.

- Employees with a spouse and dependent child OR two or more dependent children (family of three or more) will pay, on average, $4,000 per year.

- The company will also scale annual costs to the different salary levels of employees (that is, it will charge more to the employees with larger salaries).

- For all employees, the co-pay for most medical services will increase from $10 per service or office visit to $15.

Advice on Conveying the News

Imagine how you would deliver the news of the policy change, perhaps drafting your own memo. Take into account the "Negative News Pitfalls" from pages 110–12 and these more specific tips:

- Try to address the employees as valued colleagues throughout the memo. As suggested in pitfalls 1, 2, and 5, you need to strike the right tonal balance: you must show empathy while not evoking a dreadful vision of bad consequences.

- As suggested in pitfalls 9 and 10, the amount of information you supply in this memo will be critical to your colleagues' understanding the need for the change in benefits and the impact on their salaries and well-being. As you figure out what information to include, consider the following questions: What would a typical employee know, or not know, about medical-benefits costs? Will employees suspect that the company's new strategy is just the beginning of the end? How can you reassure them while also making clear that economic necessity will require their sharing some of the benefits costs?

- Look closely at the financial numbers provided under "The Basic Facts about the Benefits Changes." Also, note that the company will continue to absorb about 80 percent of the benefits costs.

- There can be no obfuscating in this communication, no fabricating reasons that can easily be refuted (see pitfall 4). In this scenario, your company is actually productive and profitable, but the rapid increase in benefits costs constitutes a serious drag on the business's ongoing prosperity.

- With reference to the "brighter future" discussed in pitfall 12, don't be tempted to promise to give back to the employees (in salary increases, for example) what you are trying to save the company in benefits costs. And don't imply that the benefits might be increased sometime in an improved financial future — that might never occur.

- As pitfall 8 suggests, you should make clear that you are announcing a company *decision*, not just exploring options and possibilities. Do not postpone the bad news and the supporting details to a future meeting or conversation. Your task is to present the bad news, tactfully and persuasively.

- The suggestions of pitfall 11 are especially important at the time of a major company decision of this sort: a well-run company will offer follow-up Q&A sessions and probably further memoranda with a greater level of detail. So consider announcing a follow-up meeting at which people can ask questions and express concerns. Whom might you bring to this meeting for expert perspectives? To whom can employees address questions and concerns before the meeting takes place?

An Effective Memo about the Benefits Changes

Here is a successful student response to this complicated scenario:

First Software Inc.

MEMORANDUM

DATE: June 2, 2016
TO: All Employees
FROM: Andreas Nitsche, Chief Executive Officer
Subject: Medical Benefits

Thanks to the dedication and hard work of all employees, First Software Inc. has managed to stay competitive and profitable for many years, and it expects to do so in the future. For the past 20 years, the company has provided all employees with a generous health-care package that far surpasses that of other companies in this community. The health of our employees and their families has been, and always will be, an important tenet of our business philosophy.

As many of you know, the cost of providing these benefits has risen rapidly over time. In the last three years, our finance department has observed an increase of 47 percent in the cost of the company's medical plan: from an average annual cost of $8,500 per employee in 2012 to $12,500 per employee in 2015. This rapid increase translates into an additional expense to the company of $1.2 million per year, or a total of $3.75 million over the last three years.

Because of these increases, our company finds itself in a situation where, in order to stay competitive and successful, it cannot continue to bear the full cost of our employees' health-care package. I sincerely regret to inform you that, after long and careful deliberation with our finance specialists, our company has decided that it will have to share some of these costs with its employees. I can assure you that after 20 years of providing full payment for the health-care package, this decision was neither arbitrary nor easy for us to make.

Beginning August 1, 2016, we plan to pass on 20 percent of the cost increase to all employees. During the last few months, our finance department worked out a formula for determining each individual's annual share of his or her health-care package. Each employee's contribution to the medical-benefits plan will be based on income level, marital status, and number of dependents. An estimation of an individual's or a family's annual payment is provided below.

- $2,000 for each individual employee
- $3,000 for each employee with a family of two (whether spouse or child)
- $4,000 for each employee with a family of three or more
- $5 co-payment increase for most services, per service or office visit (all employees)
- The contribution will also depend on the individual's level of income (i.e., lower-income earners will pay less than high-income earners).

This plan asks a lot of you, but, again, I can assure you that we believe these steps are necessary and in the best interest of the continuing success of our company. To emphasize how important this matter is, the Human Resources Department, together with our Accounting and Finance departments, plans to hold three meetings in the company's conference room on Friday June 3, Wednesday June 8, and Friday June 10, from 3 p.m. to 5 p.m. Present will be Ms. Miller from the HR Department, Mr. Hersh from the Accounting Department, and Ms. Smith from the Finance Department. These meetings will give you the chance to ask questions or express your concerns regarding the upcoming changes. I will be present at all three meetings to accept any questions you might want to ask me in person. Also, if you have any questions regarding your benefits plan right now, please feel free to contact Ms. Miller from the HR Department at 805-123-4567 for immediate response.

The long success of First Software Inc. is based on the dedication and hard work of all the employees. The decision concerning the extra payment we are asking you to contribute was made after long deliberation and, as I mentioned earlier, was not made lightly. Providing our employees with a good health-benefit plan will always be one of our major priorities. I ask for your understanding of the steps we have to take in order to keep First Software Inc. a thriving and successful company.

Sincerely,

Andreas Nitsche
Chief Executive Officer, First Software Inc.

Applying What You've Learned

Considering the advice and examples presented earlier in this chapter, and the suggestions provided below, work through the following activities, which focus on presenting company policies and procedures effectively to colleagues.

APPLICATION 5-A

Revise an Off-Putting Request for a Promotion

The background. Making personal or professional requests to your supervisor can be tricky. You might, for example, need to request time off for a family emergency, to ask for a further review of a company practice that adversely affects your area of responsibility, or to make a suggestion for improving decision making within the office.

In this application, you have been a paralegal at the law firm Quincy & Hale for about one year. During the interview for the job, it was mentioned that promotions and raises are generally considered after six months of employment at the firm. You received a positive review at the end of your three-month probationary period, but after nearly nine more months on the job you have heard nothing about the possibility of a promotion or a salary increase. Below is a letter to your immediate supervisor that does not use the best strategy to gain her support. Can you improve the tone of this letter and make a more appealing argument to your supervisor?

Dear Ms. Jackson:

I've been here at Quincy & Hale for almost a year and got a positive review after my first three months on the job, so I am wondering whether I could have a raise and a promotion. When I was hired at the firm, I was told that this was the common practice after six months of employment. Could we discuss my options? Thanks.

Sincerely,

[Your name]

The purpose. Your goal is to gain serious consideration from your supervisor without sounding unappreciative, presumptuous, or greedy.

The audience. This supervisor receives many individual requests from her colleagues. She is likely to respond more favorably to a request that is based on evidence and logic than to a communication that sounds more like a personal complaint.

The communication strategy. You don't want to sound like a child reminding a parent, "You promised." Rather, present your case in a respectful and fact-centered manner, and convey your dedication to the company regardless of the result of your request. No one likes an ultimatum.

APPLICATION 5-B

Request Information about Office Supplies

The background. Especially in the early stages of your career, you might not have a lot of authority. You will still need, however, to make things happen that are related to your responsibilities, and at times you might be asked by your supervisor to gather information and perspectives for her or his use in reaching a decision.

In this application, you are an administrative assistant at Techno Inc., and one of your responsibilities is to watch over the purchase and distribution of office supplies for 12 units within the company.

Your supervisor is concerned that there is insufficient accountability for the rapidly growing costs of office supplies (for example, stationery, mailing envelopes of various sizes, copy paper, toner cartridges, pens and pencils, and sticky notes). She wants you to write a memo to the unit directors asking each of them about their anticipated volume of office-supply use for the coming few months.

The purpose. The central goal of gathering this information is to help your supervisor project reasonable costs within the company's annual budgets. She would also like you to suggest in the memo that all employees become more aware of the costs of office supplies and attempt a more economical use of them.

The audience. Some of the directors will not like the micromanaging implications of this inquiry and might not want to take time away from their regular duties to gather the information that you are requesting. At the same time, they will see that you are writing with the authority of your supervisor, a factor that should work in your favor.

The communication strategy. Consider how you will word your inquiry to gain an accurate picture of the anticipated use and costs of office supplies. At the same time, try to establish a reassuring tone, one that avoids irritating the directors and their staff and encourages cooperation with your request.

APPLICATION 5-C

Revise a Poor Communication about Office Space

The background. Very often an organization needs to communicate news that benefits some people and has a negative impact on others, or news that is a "mixed blessing" for all concerned. Writing successfully in such gray areas requires a great deal of tact.

In this application, you are the director of finance at Caboodle Industries and need to inform a recently promoted colleague that the private office space he had expected to receive is not yet available (but probably will be in the next six months). This is a case of good news tempered by disappointing news. You will see that the first-draft message to your colleague, shown below, does not do a good job of conveying the negative news or of making clear how the issue will be resolved. Can you do better?

Dear James:

I am very pleased by your recent promotion to accounting associate. We had hoped that an enclosed office would be available for your new level of responsibilities, but that's not yet in the cards. Sorry about that, but we will keep looking for a new space. I wish you all the best as you tackle your new tasks.

[Your name], Director of Finance
Caboodle Industries

The purpose. You want to at least temper James's disappointment with the positive news that he will soon receive the promised office space.

The audience. Let's assume for this exercise that James is a fairly reasonable person, one who can accept postponed gratification — as long as he sees evidence of a better result soon to come.

The communication strategy. Don't sugarcoat the bad part of this news, for that could seem patronizing, but do show your understanding of *why* James needs the enclosed office for his higher level of accounting duties. Be as specific as you can be about the reason for the delay, and offer a realistic estimate of when the new office will be available. (The poorly crafted memo above sounds like a vague and an empty promise.) You will probably want to close this memo or e-mail by thanking James for his patience while the new office space is created.

APPLICATION 5-D

Revise a Poor Communication about Child-Care Policies

The background. It is not uncommon for businesses to announce cutbacks of various types: for example, in available overtime hours, in medical benefits, or in services previously provided to some percentage of the staff. Presenting negative news in a manner that does not diminish the trust or work commitments of colleagues is an essential skill for successful leaders.

In this application, Urkel United is announcing a policy that reduces child-care services and facilities, a policy described in the following letter. Can you rewrite the letter to reflect greater empathy (and a more sensible implementation time line)?

Dear Colleagues:

As you know, our company is facing hard times, and we regret to inform you that Urkel United can no longer afford to offer free day-care services to *all* employees with children from ages one year to four years old.

Employees who have used work-site child care for one full calendar year or longer may continue to enroll their eligible children in the Urkel facilities but at a charge of $50/day. All other employees, who have either never used our day-care facilities or who have been employed at Urkel for less than one calendar year, will not be permitted to use the facilities in the future.

These policies will go into effect 30 days from the date of this notice.

Please contact me at ext. 8663 if you would like to discuss alternative day-care arrangements. We look forward to your cooperation in implementing this policy. As always, we value the work you do for Urkel United. As the holidays approach, we wish you and your family all the best cheer of the season.

Sincerely,

Bonnie Voyage
Benefits Coordinator
Human Resources

The purpose. Your difficult task is to be candid about the reasons behind the bad news and to show that you understand the serious impact that the reduced or eliminated support for child care will have on many of the employees. For example, consider that $50/day in child-care costs will total around $1,000 each month, and employees who are no longer eligible for the company-provided care facility will likely encounter even higher costs and face logistical difficulties.

The audience. Assume that no one will welcome this news, and some will want to "kill the messenger," as the saying goes. Also, because child care for working parents has become such a serious workplace issue, even staff without children may be unhappy with this news. Given these likely reactions, you will want your colleagues to trust that the reduction is necessary, that you and others have examined all alternatives, and that you have a compassionate concern for employees and their families.

The communication strategy. Revisit the "Negative News Pitfalls" on pages 110–12 of this chapter. Given the nature of this communication, consider the crucial importance of tone and of evidence that explains why the policy change is necessary. The letter should also identify a person to whom affected employees can turn to express their concerns and to learn about alternative child-care resources. Many well-run companies would also schedule one or more open meetings to address employee questions and concerns.

APPLICATION 5-E

Revise a Poor Communication about Flexible Work Schedules

The background. Some business communications amount to "feasibility studies": their purpose is to gather information so that managers can make informed decisions about possible new policies or benefits. For example, a company might want to explore the possibility of making company-leased automobiles available to more employees or of creating a fitness facility. In such communications, you should make clear that nothing is being promised; rather, you are simply gathering colleagues' views on the issue.

In this application, United Services is exploring the possibility of allowing some number of flexible work schedules for employees. The following inquiry is poorly written and shows little understanding of the strategic issues at stake for United. Can you think through the good and bad consequences of a flexible work schedule and craft a clearer and more realistic set of suggestions and questions?

DATE: January 18, 2016
TO: All Staff Members
FROM: Judy Stabler, Chief Executive Officer
           United Services Inc.
Subject: Flex Work Schedules

The bosses and I have been considering flex-time possibilities — you know, where you have more choices about your daily time commitments. (BTW, this does not give anyone the opportunity for slacking off.) In order to accommodate your various lifestyles, we want to hear what would work for the type of work you do and for your personal interests.

Would you, for example, be interested in any of the following:

- Work from home one day each week?
- Work four 10-hour days each week?
- Maybe you could work 10 a.m. to 7 p.m. instead of 8 a.m. to 5 p.m.?
- Go half-time and split your work with someone else?
- Do you want to work a "9/80" schedule; that is, work 80 hrs in 9 days and have a long weekend every other week?
- Work five days each week but 10 a.m. to 7 p.m.?
- Do the regular schedule every other month, and flex schedule the other months?

Let us know what you think and we will devise a plan of action. Tell us why a particular flex plan would work for your job responsibilities. Until we have approved a new plan, of course, everyone will work as usual.

The purpose. Your challenge is to gather information in order to assess both what types of flexibility the staff might desire and (very important as well) whether the company can both support flexible schedules and sustain its productivity. You don't just want to hear your colleagues' enthusiasm and personal schedule desires; instead, you need their realistic assessment of how the work of the company can get done expeditiously and seamlessly in a flex-schedule environment.

The audience. Given the increasing popularity of flexible scheduling and of the ability to work from home, much of your audience will be receptive to your inquiry — and excited about the possibilities it suggests. However, some employees may wonder how important office interactions will be coordinated if flexible scheduling takes effect.

The communication strategy. You will quickly perceive that the sample memo above is poorly crafted in tone and writing style. You will want, for example, to eliminate the hierarchical reference to "bosses" and to simplify the confusing and repetitive scheduling suggestions. Be sure to stress that nothing is being promised regarding future schedules; rather, you are seeking useful insights for the decisions still to be made. To aid that assessment, what sorts of practical, work-related information might you need from the various office units?

APPLICATION 5-F

Revise a Poor Communication about Holiday Office Coverage

The background. Sometimes special circumstances, such as major holidays or emergency situations, require a temporary shift in work schedules or in the availability of certain company resources. Some companies develop "what if" work protocols before unusual events even occur; others have to improvise when a sudden need arises.

In this application, Framish Fiduciary needs work coverage over the Thanksgiving holiday, and no plan for this coverage is in place. The HR director has, however, written a memo that will probably irritate employees on every level. For one thing, the memo lacks empathy. For another, it is confusing: is the HR director announcing a plan, requesting input, or both?

Review the current memo and then revise it so that it makes a clear announcement of a settled plan.

Dear Colleagues:

As we approach the Thanksgiving holiday (Thursday and Friday, November 24–25), we need to plan for coverage of essential functions during those days. All nonessential staff will take November 24 and 25 off as scheduled.

The situation is different for our security guards, for the computer technicians who maintain our Web site services, for the phone receptionists (needed on November 24 only), and for all staff working on the Behemoth Project (for which the completion deadline is Monday, November 28). All employees in these categories will need to work regular hours, in shifts to be arranged, on either November 24 or 25 — or perhaps both days.

Each division manager will submit a holiday-coverage plan for approval by November 7. Employees who are required to work on November 24 and/or 25 will receive compensatory vacation day(s) to be taken in the month of December 2016. Thanks for your cooperation as we all pull together to make Framish Fiduciary Inc. the most reliable financial-service agency in New York.

Sincerely,

James Kiehl
Human Resources

The purpose and audience. Again, the goal is to make the memo an announcement of a settled plan for the holidays (which you might need to flesh out a bit), and *not* also a request for the directors to submit further information. You will want your office-coverage goals to be clear, reasonable, and sympathetic to those who must work through Thanksgiving.

The communication strategy. Even though such terminology is common in business organizations, no one wants to be regarded as "nonessential." However, few if any employees will enjoy working through the holidays, even if it means two extra days off in December. See if you can clarify the responsibilities needing coverage during the holidays and write, in general, with a more sympathetic understanding of the (temporary) hardships to be endured by some colleagues.

APPLICATION 5-G

Revise a Poor Communication about Employee Parking

The background. All professional communications should embrace a clear thought process, and this is especially important when a new strategy or necessity is being announced. Effective leaders must always demonstrate that they have thought carefully about the practical consequences of revising policies and strategies.

In the following application, K. Smedley, an assistant director at Jackson Architectural, is trying to suggest alternatives to an overcrowded company parking garage. However, Smedley has not thought through the parking remedies carefully; moreover, she or he has written the recommendations in a confusing manner. Help Smedley develop a more sensible set of strategies, a clearer presentation of them, and a more sympathetic tone.

Dear Colleagues:

All of us face frustrations over the poorly designed, limited parking facilities at Jackson Architectural, Inc. The management has determined a new plan that will alleviate many of these concerns. Every Monday, a day that generally brings fewer clients to our firm, staff of Jackson Architectural may park in the spaces designated for "visitor parking" when these spots are available. On Thursdays (and on other days, for that matter), we encourage you to carpool, to ride the #12 bus to work, or to find parking on the main street (but note the two-hour limit). On Tuesdays and Fridays, for a small fee, employees may park in the furthest section of the lot owned by our neighbor Forthright Towing.

We are all in this together, and I regret the inconvenience.

K. Smedley
Assistant Director of Business Services

The purpose. You need to make a compelling case for remedying the parking situation and to devise realistic parking and transportation strategies that are less obtuse and insulting. To accomplish these goals, you will need to imagine your way inside this business situation.

The audience. Nearly all working people are concerned about their commuting time and, if they drive to work, the ease with which they can park their car. So this is a sensitive topic for your audience.

The communication strategy. As previously suggested, you must re-imagine more realistic parking options for your valued colleagues, explain the options clearly, and demonstrate your understanding of the frustrations that even a well-conceived plan will cause for some of your colleagues.

CHAPTER 6

Writing to External Constituencies

Understanding the Challenges of Writing to External Audiences

Assuming that you start out in a relatively low position in a company's hierarchy, it might seem as though your role is fairly inconsequential, that you have little impact on the organization's reputation or growth. However, external audiences with whom you communicate will view you as a representative of the company: you will help shape these audiences' opinions not just of you but also of your organization as a whole. It's a big responsibility and not one you should take lightly.

Your handling of a seemingly trivial business interaction can have greater ramifications for your future and that of your company than you might realize. You can never be certain, for instance, when your careful management of an unhappy customer might turn her into a loyal one. Also, a person seeking company information today might eventually become an important investor or donor, and a client of modest financial standing might later become a major force in the market. It's important, then, to treat every external audience with courtesy and respect.

Most of the strategies discussed in Chapter 5, "Writing to Colleagues within the Organization," apply to communications with those outside the organization as well. The same qualities of clarity, accuracy, tactfulness, and respect should shape all business writing, whether the audience is internal or external. Clients, customers, suppliers, service providers, and professional colleagues at other companies are the lifeblood of any organization, and your ability to communicate respectfully, responsively, and forthrightly with these constituencies will contribute greatly to your, and your employer's, success.

Knowing That Word Gets Around

In the twenty-first century, it's much harder to count on the privacy of personal information or communications than at any time before the information revolution fostered by the Internet. There has also been a concurrent movement, especially in the so-called Generations X and Y (who are the primary audience for this book), to place much less value on personal privacy. Compared with their parents and grandparents, many members of the current generation are much more candid and open concerning their personal (social, romantic, medical, and financial) information. They can also reach far wider audiences for their communications, through texting, tweeting, blogging, and social networking.

The Pluses and Minuses of More Open Communications

This evolution toward more self-expression and transparency certainly has its virtues in both personal and business communications. Businesses can, for example, easily mine available information, whether wireless or wired, to interact directly with customers, to personalize shopping experiences, and to address customer concerns with greater immediacy.

A potential drawback of this information revolution, from a business standpoint, is that whenever we write to just one person, we are potentially writing to the world by way of the Internet. For example, a customer who receives a rude response by e-mail to a complaint about a company's product or service might be tempted to forward the response to friends or colleagues or to post it to a consumer Web site, blog, or Facebook page. Thus, the impact of one rude message is multiplied, with potentially serious consequences both for the sender of the message and for the organization that he or she represents. (For more examples of this multiplication effect, see the CEO's e-mail threat on page 106 and the mistaken college admissions offers on page 228.) In addition, businesspeople whose e-mails, online postings, or other electronic communications include remarks that are sexist or otherwise discriminatory or offensive, or that use irresponsible or illegal strategies, can imperil not only their professional standing but also, by extension, the reputation of their company.

Legal Implications of Problematic Communications

On an even more serious note, you should be aware that very few business communications can be protected as "private" when a legal challenge is set in motion. Hardly any past communications are entirely "off the record" when attorneys and courts pursue the evidence in a case, and many an erased or expunged electronic communication can be recovered by experts. The dramatic cases made against the top managers of Enron in 2001 and of Tyco and WorldCom in 2002 provide ample evidence of the seemingly private becoming public—and actionable. All three cases were extraordinarily complex, but in each one many thousands of supposedly private e-mails and other company documents came to light, exposing fraudulent financial claims and crooked dealings. These communications provided prosecutors with ammunition to pursue cases against the CEOs of these companies and a number of highly placed managers.

As these examples make clear, bad business behavior is increasingly easy to discover, publicize, and punish. Therefore, in all of your business communications your commitment to fairness, honesty, and respect is as pragmatically necessary as it is personally admirable.

Principles of Respectful Communications

With these goals in the background, the business writer communicating with external constituencies should observe these principles:

- Write to others in a clear-minded and respectful way.

- Do not respond in kind if your correspondent is being flip, sarcastic, or disrespectful. Remain professional and respond calmly, with patience and understanding. Be careful as well not to patronize or talk down to an irate customer or client. Be sympathetic but also pragmatic in your response to complaints.

- Refrain from sharing personal information or perspectives in your business communications.

- Familiarize yourself with the policies and guidelines that govern the business and communication practices of the company you represent. This preparation will help you follow such standards readily in your daily correspondence.

- Take for granted that there can always be an unintended audience for any communication. Write only what you would be prepared to explain, justify, and successfully defend.

Providing Information Clearly and Persuasively

Many communications with external constituencies are primarily informational. Here are just a few examples:

- Your company is seeking external bids on a new project or product development. You are responsible for issuing the guidelines and specifications and inviting targeted organizations to respond with the necessary information.

- You need to write letters to new hires, letters that are affirmative in tone and that state the contractual aspects of the job offer: for example, the salary, benefits, starting date, and personnel review process.

- Your company is proposing to share warehouse resources with another company for mutual benefit, and you must outline the details of the proposed arrangement.

- You have been asked by a potential investor for an annual report and must write a polite and persuasive letter to accompany it.

An Effective Response to an Information Request

Here is how a CEO might respond in a personal and effective way to a potential investor's request for information about his company:

Foreground Industries
6615 82nd Street
New York, NY 10019

Mr. Benito Tanaka, CEO
Timely Enterprises
1215 First Avenue
Albany, NY 12208

January 19, 2016

Dear Mr. Tanaka:

We are very pleased that you continue to be interested in our research on a new malware protection system for home computers. Over the past year, Foreground Industries has invested $255,000 in the development and testing of this product, and an additional six months and further funding of approximately $100,000 will allow us to bring the product to market.

As you requested, we have enclosed our Annual Report for 2015 and additional specifications regarding our malware product, "Killroy." We hope that these materials will encourage you to explore an investment opportunity with Foreground Industries. Our product operates on a technology different from, and 40 percent more effective than, the approach used by currently available malware protection systems. We are therefore confident that over the next two years we can gain a significant market share in the malware field.

If you find our product research interesting, we invite you to meet with our chief software engineer, Mary Adams, and our CFO, Dave Grazer, to discuss possible terms for an investment in Foreground Industries as we move forward with Killroy.

We appreciate your interest and consideration. If you have further questions, don't hesitate to contact me by e-mail at jfeldstein@foreground.com or by phone at 815-666-2318.

Sincerely,

Jonathan Feldstein, CEO

In this cordial interaction, the CEO of Foreground Industries writes in a friendly and respectful manner, providing preliminary information to the potential investor, Benito Tanaka. Feldstein also includes a brief time line for the new product's development, the investment amount desired, and an offer to have experts from Foreground Industries meet with Tanaka to discuss the new product and the potential returns on investment.

An Ineffective Alternative

Feldstein could have made his letter much more generic and brief:

Dear Mr. Tanaka:

Enclosed please find the material you requested regarding our Killroy malware protection systems and the investment opportunities. We look forward to talking with you further.

Instead, he included specifics that lent a more persuasive edge to an otherwise routine letter. He also took the opportunity to treat the recipient and potential investor, Mr. Tanaka, as someone special—a potential partner in a shared enterprise. Mr. Tanaka is therefore more likely to want to conduct business with Mr. Feldstein and his company.

Keeping the Human Touch in Big-Business Communications

Whether large or small, businesses sometimes resort to templates like the following:

> Thanks for your inquiry. We will give your interests all due attention.

Such vague, generic communications are common, and you may have seen them in response to a job inquiry, to a complaint, or to a customer-service request. But no one likes receiving these formulaic communications; it's difficult to believe that there is any actual human agent behind them, and it may seem unlikely that the company will take any action in response to the applicant, customer, or client.

In contrast to ineffective responses like these, large, well-run companies that rely on mass e-mail communications with thousands of customers take considerable time to craft these messages with a human touch. Consider, for example, this typical message, computer-generated and disseminated, from Amazon:

> Hello Jane Smith,
>
> Thank you for shopping with us. We thought you'd like to know that we shipped your items, and that this completes your order. Your order is on its way and can no longer be changed. If you need to return an item from this shipment or manage other orders, please visit *Your Orders* on Amazon.com.
>
> Your package is being shipped by USPS, and the tracking number is 9331997642074730692605. Depending on the shipping speed you chose, it may take 24 hours for your tracking number to return any information.
>
> Returns are easy. Visit our *Online Return Center*.
>
> If you need further assistance with your order, please visit *Customer Service*.
>
> Amazon

Amazon has taken care to construct a clear, relevant, friendly message, and the company has embedded links (*Your Orders, Online Return Center,* and *Customer Service*) relevant to this particular customer, Jane Smith. Whether you are one person writing to one recipient, or a committee writing (by way of a computer algorithm) to thousands of clients or customers, you should take the same care to create messages that are actually useful to and respectful of your recipient(s). That level of concern in communicating helps create and sustain a respected company profile and good customer relationships.

Avoiding Business Liabilities

As suggested earlier in this chapter and in Chapter 5, you as a business writer must always be aware of and alert to the potential negative impact of using an offensive word or phrase, introducing an unintended ambiguity, seeming to say something discriminatory or dismissive, or promising more than you can reasonably deliver.

Concerns over business vulnerabilities and legal liabilities are intensified when you write to external constituencies. You need to learn, for example, how to apologize for a company error without encouraging an ongoing dispute or providing too much information, which could serve as ammunition for further fault-finding. Conversely, providing too little information could come across as flippant, dismissive, or even evasive. You need to learn just how much information makes a reasonable case for reasonable people.

Early in your business career you will need to think consciously about such issues when you communicate. You will also need to become familiar with your organization's policies and guidelines for communicating with clients. In some companies, these expectations, or protocols, are articulated in a formal manual; in many organizations, however, the expectations are a less formal legacy that can be learned only by asking more senior associates (especially your supervisor). As you gain professional experience, you will begin to easily recognize the pitfalls and land mines of business communications, and avoiding such problems will start to become automatic. Until you gain this level of experience, double-check with your supervisors that you are communicating effectively with clients. Your supervisors will appreciate that you are being proactive about making a good impression on the company's behalf. Keep a variety of audiences in mind as you write, not only the specific recipient(s) to whom you are communicating but also the much larger communities of colleagues, clients, or competitors to whom you might need to explain and defend your writing choices. That exercise of *imagining your way inside* the current situation and simultaneously anticipating its possible broader implications will prevent most errors of consequence.

Responding to Real-World Writing Scenarios

In the next section, we'll examine how two students responded successfully to a typical—and challenging—situation involving an external audience: a customer complaint. Then, you will apply what you've learned from this scenario and from other explanations and examples in this chapter.

Analyzing a Writing Scenario: Responding to a Customer Complaint

Along with generating myriad informational and largely positive communications with external constituencies, professional writers also encounter requests that can't be fulfilled, challenges from unhappy clients or external associates, and even threats of legal action. For a broad sample of what customers and other reviewers write about products, services, and entire companies, take a look at reviews posted on the Web sites of retailers like Sears, Home Depot, or Macy's, or on such review sites as yelp.com, buzzillions.com, reviews.cnet.com, and consumersearch.com. See also the ratings of the best and worst customer service on customerservicescoreboard.com. These sites remind us that *everyone with access to a computer can publish a positive or negative opinion for the world to see.* At the very least, you want most of your reviews and comments to be positive. Dealing effectively with customer complaints is one significant step toward achieving and maintaining a favorable reputation.

The Complaint

Here is an example of a customer complaint that can probably be addressed successfully, but only with great tact and strategic thinking on the respondent's part:

Dear Sirs:

I have been buying from Framistan Technologies for my company for 10 years, and I must say the last shipment was badly handled. First, I understand from your discourteous shipping director that there was a delay of three weeks before my order was even processed. Then the shipment that arrived was poorly packaged. About 20 percent of the motherboards were damaged in transit. The number of units shipped was also 12 items shy of the total 100 requested.

This is really bad service, especially for a longtime customer like me. I demand an explanation of what happened, your assurance that the order-processing problems have been corrected, and immediate completion of my order. This needs to happen ASAP, for I have my own customers waiting. I will expect to hear from you right away.

Miriam Tiger, President
Calcified Industries

Let's assume the following in this complaint-resolution scenario:

- Miriam Tiger is an important customer, and your company did a poor job in processing her order.

- You need to apologize for the problems she experienced, briefly explain the reasons for the problems and how they have been or will be resolved, and get the missing products into her hands as fast as possible and the damaged products replaced as quickly as possible.

- You might also consider a current or future discount for this important customer.

- You must somehow take all of these steps while still standing behind your staff (she complained about your shipping director)—or at least protecting the confidentiality of personnel actions at your company—and without making your company vulnerable to further attack.

One Effective Response to the Complaint

Here is how one student, Michael Cipriano, responded to Ms. Tiger:

Dear Ms. Tiger:

First, I would like to sincerely apologize for the mishandling of your order and for the poor condition of the merchandise. Over the past several weeks, we have been experiencing software problems with our shipping process, and we have worked around the clock to get it fixed. At Framistan Technologies, we pride ourselves on quality products and service, and I hope that you would agree that the mishandling of your order is not characteristic of the service you have received in the past.

You will, of course, be issued a prompt refund for the damaged items; the missing units from your shipment, along with replacements for the damaged items, have already been shipped via FedEx Priority Overnight shipping. You should expect the shipment no later than 3 p.m. tomorrow, April 13.

I would also like to apologize for the unacceptable interaction with our shipping director. We are looking into the incident and will take the appropriate steps to ensure that it does not happen again. Thank you for directing our attention to this matter. If you have any more questions regarding your order, or any future order, please do not hesitate to call me directly at 805-292-6435 or to e-mail me at cipriano@relations.framistan .com. We hope that the shipment arriving tomorrow will complete the order to your liking and that you will continue to be a valued customer of Framistan Technologies.

Sincerely,

Michael Cipriano, Customer-Relations Manager
Framistan Technologies

Notice the candor with which Michael Cipriano addresses the mistakes made by Framistan and suggests how they were resolved. Yet he does not get into details about the incident with the shipping director. Many laws and statutes protect the privacy of personnel records, so it's best for you to divulge very little in this area should you have to address an incident like this. In addition, you should support your colleagues whenever possible. Before you accept blame on an employee's or on the company's behalf, be certain about precisely what transpired.

Cipriano is also very clear about when and how Ms. Tiger's order will be completed, understanding that her own customers are waiting. In addition, he has decided to refund her for the damaged items as well as ship new ones, thus providing a significant discount on this order. One of this customer-relations manager's best strategies is the reminder that his company usually provides excellent service and his hope that Ms. Tiger will agree with this assessment: "At Framistan Technologies, we pride ourselves on quality products and service, and I hope that you would agree that the mishandling of your order is not characteristic of the service you have received in the past." This sentence calls to mind Miriam Tiger's primarily positive experiences doing business with Framistan.

Another Effective Response to the Complaint

Here is a somewhat more concise response by another student:

Dear Ms. Tiger:

You are one of our most valued customers, and I would like to personally apologize for the poor service you received from Framistan Technologies. I have looked carefully into this matter, and I can assure you that we are taking all the steps necessary to resolve this problem.

To take full responsibility for our mistake, Framistan will reimburse Calcified Industries the entire amount of the last shipment (invoice # 2458346). We will also deliver a replacement of the damaged motherboards and the missing items. A new shipment has been sent out via FedEx Priority Overnight and should arrive tomorrow no later than 3 p.m. In addition, I would also like to offer you free delivery on all purchases for the next six months.

I sincerely apologize for this inconvenience and hope that our partnership will continue to grow despite this matter. If you have any questions or concerns, please do not hesitate to contact me directly at (805) 576-1244. Thank you for your patience.

Sincerely,

Sheena Joseph, Customer-Relations Manager
Framistan Technologies

This student has written a more broadly applicable "template" letter, but it is nevertheless friendly and addresses the specific problems experienced by Miriam Tiger. The student decided not to say anything specific about the complaint against the shipping director, but she does broadly acknowledge Framistan's "poor service" and assures Ms. Tiger that "we are taking all the steps necessary to resolve this problem." The student, as customer-relations manager, has also offered a compensatory perk to Miriam Tiger, free shipping for the next six months.

Whenever you want to offer discounts or other compensations to a client, be certain first to check on company policy with an appropriate supervisor. Special customer perks, like rebates and discounts, should generally be offered to entire categories of people, not just to particular individuals, because word spreads rapidly in customer communities. For example, your company might offer discounts on large orders. In all cases, offers like these should be explainable according to defined principles.

Further, if you offer discounts or other perks, be sure to *limit your company's liability*! For example, if you offer "a 30 percent discount on your future orders" to an aggrieved party, he might take advantage and buy out your entire stock at a very attractive price—never needing to buy your product again. Thus, you must always define and limit the potential monetary impact; for example, "We would like to offer you a 30 percent discount on your next order up to $1,000." (Notice that the coupons we sometimes receive for store or restaurant discounts always have expiration dates and other defined limits on how we can use the special opportunity.) This strategy allows customers to enjoy the benefits of a discount or deal without bankrupting your company.

Applying What You've Learned

Considering the advice and examples presented earlier in this chapter, work through one or more of the following activities, which focus on writing to external constituencies. Each activity offers guidance on defining the writer's purpose, determining the needs and expectations of the audience, and making the best choices regarding tone and strategy. The final section of this chapter includes examples of how individual students crafted responses to a few of the business situations presented here.

APPLICATION 6-A

Respond to an Information Request from a Potential Investor

The background. Sometimes individuals or organizations seek information from a particular company. The request could come, for example, from a potential investor or bank lender, a government agency that regulates your industry, or one of your service providers. When you receive such requests, you should first find out from more senior colleagues whether you or someone else should respond and how much company information can be provided. Most companies want to be relatively transparent; however, they will not want to divulge confidential records or trade secrets to an external constituency if they are not required to do so.

In this application, the chief financial officer of an investment firm has heard that your company, Royal Pets, recently developed a new line of organic pet foods, which you are just beginning to place in local stores. She might be interested in investing in your product and helping to place it in a range of retail stores with which she is associated. As the marketing assistant for Royal Pets, you have been asked by the director of marketing to draft a response for her review. Can you include in the draft some basic information about your product and outline the successes you have already had with sales? (For this application, you will need to make up some plausible facts and figures and offer to meet for further discussion.) Here is the letter of inquiry:

Dear Royal Pets:

I read in our local paper *The Happening Times* about the organic foods for dogs and cats that you have created and are beginning to market locally. I have lots of experience in product placement in the retail industry and might be interested in investing with your company. If you think this sounds like a possible opportunity for you, please send me your company's product, marketing, and financial profile. We can then discuss possible further steps.

Sincerely,

Juanita Flores, CFO
Investment Cooperative

The purpose. Your draft response to the potential investor is a preliminary means of testing the waters, a way for both sides to see whether further, more serious discussions are warranted. So you will need to provide only a few key points about your new product and some basic sales information. Your company does want to attract investors; thus, the letter you write should be positive and persuasive.

The audience. Based on her letter, Juanita Flores may have seen only a newspaper article about your company. Therefore, she will probably need a more detailed description of your new organic product and why it would benefit pets and be attractive to pet owners. She will also need some sales and financial facts to determine whether Royal Pets would be a good investment opportunity for her.

The communication strategy. In a short (perhaps one-page) letter, combine your enthusiasm for the new Royal Pets product with some key facts. In other words, ground your persuasive energy in evidence.

APPLICATION 6-B

Write a Company Mission Statement

The background. Many for-profit and nonprofit organizations develop succinct mission statements that both advertise their goals and commitments to a broad external community and help the employees agree on and affirm the organization's core values. Mission statements often appear on company Web sites and in annual reports, and they are sometimes echoed in advertisements. They typically emerge through a committee process, but it's also common to have one or two people draft versions of the mission as it emerges through a group process. In this application, you are part of a task force charged with producing a mission statement for Fortress Security Systems, which manufactures, installs, and services home security systems in the Dallas/Fort Worth area. After a number of discussions about the company's mission and how to describe it, you and the other task-force members agreed that you will draft the one-paragraph statement that will be suitable for the group's review.

Here are a few sample mission statements for your guidance. To look up more real-world examples, conduct a Web search for any company name and the phrase "mission statement."

Microsoft Mission Statement

Empower every person and every organization on the planet to achieve more.

National Autism Association Mission Statement

The mission of the National Autism Association is to respond to the most urgent needs of the autism community, providing real help and hope so that all affected can reach their full potential.

Wikimedia Mission Statement

The mission of the Wikimedia Foundation is to empower and engage people around the world to collect and develop educational content under a free license or in the public domain, and to disseminate it effectively and globally.

In collaboration with a network of chapters, the Foundation provides the essential infrastructure and an organizational framework for the support and development of multilingual wiki projects and other endeavors which serve this mission. The Foundation will make and keep useful information from its projects available on the Internet free of charge, in perpetuity.

Twitter Mission Statement

To give everyone the power to create and share and information instantly, without barriers.

The purpose. Fortress Security wants the mission statement to capture its interests in personal safety and security and its broader community commitments.

The audience. Fortress Security wants to be able to turn to the mission statement for employee inspiration, as an announcement of goals and principles for external constituencies, and for possible use in advertising campaigns. It's important, to begin with, that the statement embody the perspectives shared among colleagues at the company. Without that shared understanding among insiders, it will be difficult to use the statement with external audiences.

The communication strategy. This is one instance in which inspiring rhetoric is called for in professional writing. That's a tricky endeavor, for you want to offer inspiration without using vague and cloying clichés. You also want to center your rhetoric both on what's already true about the company and on what the company seeks to become. Thus, a mission statement typically balances an organization's current realities with its aspirations.

APPLICATION 6-C

Create a Return Policy for a Retail Store

The background. Typically, companies have policies on how to deal fairly not only with employees but also with customers and clients. The latter types of policies seek to avoid misunderstandings and ill feelings and also to protect a business's financial interests.

In this application, you are a new member of the management staff of SurfsUp Clothing, and you are working with three other team members to develop a return policy that customers will understand and find acceptable. After several discussions among the team members, you have volunteered to draft a policy statement for their further review.

The purpose. SurfsUp needs to clarify its return policies on beachwear and other sports attire. The company wants its customers to be satisfied with purchases and thus wants to be generous about the period of time for returns, the condition of the merchandise, and whether original receipts are required. SurfsUp Clothing also needs, however, to protect its financial interests and, if possible, prevent the abuse of a too-liberal or vague return policy.

The audience. The audience will be SurfsUp customers, who will want the return policy to be easy to understand, sensible, and fair. The return policy will be posted in the store for all customers to read and printed on all sales receipts.

The communication strategy. Begin your draft with a very brief introduction that will let your fellow managers know what you have attempted to compose. Don't hesitate to alert them to issues still unresolved (if there are any in your draft). Then present the draft of the policy and conclude by inviting your colleagues to offer their feedback.

APPLICATION 6-D

Resolve a Complaint about a Catering Fiasco

The background. Even well-run companies receive some complaints from customers and clients. It's important to respond tactfully to such complaints and, if possible, to rectify any problems caused by your company—through a refund, an exchange, a future discount, or some other compensation. A customer problem can be turned into customer appreciation when the company representative responding to a complaint takes responsibility and stands behind the product or service. The good word about this customer-company interaction can spread rapidly among other consumers, especially when customer-satisfaction Internet sites come into play.

Consider the following situation: a customer named Lia Connor hired your catering firm, Fresh Fare, to cater her daughter's sixteenth birthday party. As you will see from the following e-mail, Mrs. Connor has a number of complaints about the servers and the quality of the food. As the customer-relations representative for Fresh Fare, you have already talked with the staff member who oversaw the event, and she confirms the problems that Mrs. Connor describes. You need to apologize for the problems and somehow make amends.

Here is Mrs. Connor's complaint:

Fresh Fare Managers:

I have used your catering services several times in the past and was pleased with what you provided for my events. But the service you provided at my daughter's sixteenth birthday on September 16 was really unacceptable. The burgers took a lot longer to prepare and serve than expected (a delay of over an hour), the cake was chocolate when we had ordered a white cake with chocolate frosting, and my daughter Carli's name was misspelled as "Curly" on the cake. Two of the five staff you sent to us were dressed in cutoffs and flip-flops (while other staff were in professional black-and-white attire), and these two staff had no idea how to serve food politely and efficiently.

I am very disappointed in all of this, as is my daughter, whose important day was certainly not as special as it should have been. I think you owe us some compensation.

Mrs. Lia Connor

The purpose. Your main goal is to apologize and compensate in a manner that might retain Mrs. Connor's business in the future and dissuade her from speaking poorly of your company among her friends (or from posting her complaints online).

The audience. In this case, the complainant is pretty accurate in her descriptions of what went wrong and has ample reason to be unhappy. As the customer-relations rep, you need to extend a convincing apology, explain how things went wrong, and offer some form of compensation to Mrs. Connor.

The communication strategy. Apologizing effectively in business is an art. The apology must be sincere while avoiding hand-wringing or self-flagellating. See if you can balance these issues of tone and strategy in order to appease, perhaps even to please, Mrs. Connor.

APPLICATION 6-E

Resolve a Complaint about Customer Service

The background. While it requires great tact to respond to legitimate customer complaints (as in Application 6-D), it can be even more challenging to respond to complaints that are unfounded, particularly if you want to retain the business of the unhappy customer. In such situations, you will need to explain to the complainant that the fault lies elsewhere but without sounding defensive or pointing fingers.

This application will call for that type of balancing act. It concerns Charles Jameson, president of Rambo Corporation, who is a very important customer but who also routinely complains about poor service, even when your company has done a good job. That being said, you want to retain his business. In the current case, the stamping machines shipped to Jameson by your company, Fescue Ltd., had been set up improperly by Jameson's own crew, and your engineers quickly resolved the problems. Can you respond to Jameson's complaint in a way that is both inoffensive and candid about the issues? Here is Jameson's e-mail:

People at Fescue Ltd. —

My last order with your company was very poorly processed and resulted in some serious slowdowns in my manufacturing process. You need to be shipping me fully functional stamping machines and by the necessary deadlines in order for me to continue business with you.

I have been an important customer for many years and require an immediate response to this intolerable situation.

Sincerely,

Charles Jameson, President
Rambo Corporation
Dearborn, Michigan

The purpose and audience. Jameson is clearly angry, and, as noted earlier, he is a constant complainer as well. To make matters worse, his own employees were actually at fault for the problems he encountered with the machines you delivered to him. Your challenge will be to make clear how the problems actually occurred and to explain that your engineers have already corrected the installation errors of Jameson's workers—all the while without seeming to argue with Jameson or to shift the blame onto him.

The communication strategy. Responding successfully to a situation this complex will require all of your non-egoistic tactfulness and grace. Perhaps you can reference the challenges of installing the latest machinery without placing blame on Jameson and his workers. You can then assure Jameson that your engineers have already visited his plant and helped install the machines properly. Again, because you are dealing with a very tough but also important customer, your strategy will need to be quite self-effacing, even though your company is not actually at fault.

Note. To see how two students responded to Application 6-E, see page 154.

APPLICATION 6-F

Revise an Angry Complaint about a Cleaning Service

The background. Businesses must try to cultivate good relationships with other businesses or service providers, for word will spread rapidly if your company is perceived as treating external groups unfairly or disrespectfully. Your poor reputation in this regard can have a serious impact on your ability to hire new employees or to enlist the aid of businesses whose services or products you need to run your own company.

In this application, Jack Flanders, the facilities manager of Hardcastle Inc., risks damaging his company's reputation because of a harshly worded e-mail complaining about another business, the current custodial service. Specifically, Flanders is fed up with the poor service being delivered by a recently hired cleaning and maintenance company. Assuming that the service is in fact inadequate, how would you rewrite the angry e-mail from Flanders to make it a more effective communication? Consider how the manager's tactless memo might hurt his company's reputation with other service providers in the area. Can you devise a strategy for discussing and possibly resolving the areas of concern with the maintenance company? Here is Flanders's poorly conceived message:

Mr. Crowther:

I don't know what you people at Aces Cleaning Crew think you are doing by not maintaining the bathrooms, hallways, and offices in our building. We complained about this last week, and nothing changed. The place is a mess, and before your company started providing cleaning and maintenance services to Hardcastle Inc., everything was spick-and-span. How soon are you going to correct this situation and do the job for which we hired you?

Jack Flanders, Facilities Manager
Hardcastle Inc.

The purpose and audience. Your goal is to write a more temperate communication and to see whether you can get your audience, Aces Cleaning Crew, to meet your company's expectations. You really don't want to start over with yet another cleaning service, and you want to treat the current group, only recently hired, fairly.

The communication strategy. Think of this not as a threat against the cleaning company but as an expression of serious concern and a desire to negotiate, if possible, an improved level of service. Consider a time and place for discussing the issue face-to-face. It may be that you will eventually need to terminate the services of the cleaning company, but first you should try a more reasoned approach to correcting the problems.

APPLICATION 6-G

Revise a Letter That Delivers Bad News Insensitively

The background. Delivering negative news is always a challenge, especially when the recipient's professional or other status is at stake. Whenever possible, you should explain the main reasons for the negative news and perhaps suggest alternatives for the recipient(s) to pursue.

In this application, you must help deliver disappointing news as sensitively as possible. The recipients of the news will be the parents of Jimmy, a Troutbeck College student who has earned a 0.76 GPA for his first semester at the college (the minimal continuation standard for that term is 1.50, with 2.00 being a "C" average). A faculty committee has reviewed all the evidence and decided that there is little hope for Jimmy to recover from his awful start (at least not at this time). Troutbeck's dean has not, however, written a letter with much sensitivity or with any context for the tough decision. Can you compose a better communication to the parents? Here is the dean's letter:

Dear Mr. and Mrs. Shipley:

We regret to inform you that your son, Jimmy, has flunked out of Troutbeck College. Since he is only a freshman, we trust that he will be able to make a fresh start elsewhere. Please note that Jimmy's belongings will need to be removed from the residence hall by this Friday, December 16. We appreciate your consideration and wish Jimmy all the best.

Sincerely,

Dean Vernon Wormer
Troutbeck College

The purpose. You need to offer a more nuanced picture of the college's expectations and the extent to which Jimmy fell short of those expectations. You also need to provide a glimpse into the review process that led to the decision. Your goal is to break the news more gently and to convince the parents that the decision was not arbitrary and that the college does care about the success and well-being of its students.

The audience. This decision will have a serious impact on both the student and his parents. The consequences could include serious financial and social losses as well as a greatly diminished opportunity for the son's educational advancement. Your improved letter needs to show genuine sensitivity to these concerns. An additional audience may well be the friends and relatives of the Shipley family, who might develop negative perceptions of the college if they learn that it communicates insensitively with students or their parents.

The communication strategy. Usually colleges and universities try to notify students themselves before providing the bad news to parents. Thus, a series of phone calls might precede a more formal letter of "disqualification," as it is sometimes called.

In your dean's letter, provide a bit of context on the review process and the standards the student needed to meet. Avoid all "flunking out" language and similar terms that would imply the son's intellectual inadequacy (which might not be the case at all). Don't make it sound as though the son's belongings will be tossed into the street; rather, offer a contact person in Residential Life who could advise the parents on move-out procedures. Finally, you might briefly describe a process for the son to reapply to Troutbeck College after earning an improved record elsewhere. Many colleges and universities offer such re-admission opportunities. Close with a further expression of regret and designate a contact person in the dean's or the advising office.

APPLICATION 6-H

Respond to a Request from a Privileged Alum

The background. Most businesses get requests for special favors. For example, an outside party might ask for confidential information, the inside track for employment, or a special discount. The business for which you work might grant some of these requests, but just as often special requests are politely denied ("I wish I could, but I can't because . . ."). If a client or customer asks for a favor, be sure to check first with your supervisor regarding company policy. Whenever a special request is granted, there is always the danger that other individuals or groups will demand the same treatment or just feel unfairly served. Sometimes there are even legal vulnerabilities in a company's unequal practices. Thus, it's always best to weigh special requests first according to their legality, and second according to an established set of policies and principles (a policy on exceptions, so to speak).

In this application, Mr. Nudgely wants his underqualified nephew to be admitted to Prestige University, of which you are president. In the following letter, Mr. Nudgely is asking you to grant a special favor by overruling the admissions director. Let's see you assume the role of the university president. How and according to what principles are you going to say "no" to this important alumnus and his family?

TO: [Your name], President
      Prestige University
      Boston, Massachusetts

Dear President [Your name]:

My nephew Herbie (Jones) is a fine young man and has thus far been unsuccessful in seeking admission to Prestige University. This is a frustrating situation in that, you will recall, three generations of our family have attended good old PU, and a denial of admission for Herbie would be a profound disappointment for our family.

We are longtime supporters of PU. My wife and I currently serve on the Parents' Council, and we have helped raise more than $3 million for the University in recent years. We don't want to be forced to reconsider our important affiliation with the University.

While Herbie may not have the most illustrious academic credentials, we believe that the year he spent working for Habitat for Humanity after high school has given him the maturity he needs to succeed in college. We ask you to talk with your admissions director, Jim Stall, who thus far has not been receptive to our concerns. We will appreciate your intervention in this matter.

Sincerely,

Dick Nudgely, CEO
Grandiose Development Corp.
Newton, Massachusetts

The purpose. Your goals are twofold: to support the admissions decision, thus upholding the quality criteria of the university, and to retain the loyalty and financial support of the angry alumnus/donor.

The audience. It is clear that Mr. Nudgely is a self-important person; he is also a person whom your development office will want to "cultivate" in coming years. He will need some complimentary stroking in your letter, a display of your gratitude for his major contributions. You must also address him as an insider to the university's interests, not as an external constituency: as a third-generation alumnus, he feels that he and his family *are* the university, not mere onlookers. Given Nudgely's expectations and his importance to the university, you should take the time to sketch future possibilities for his nephew as you say "no" to the immediate request.

The communication strategy. Be sure to thank and (briefly) flatter the Nudgely family in your opening, and conclude your response with another note of gratitude. In replying to Nudgely, you need to reference the careful admission review process, bearing in mind that you probably do not have the power to change an admissions decision, and you also want to support admissions colleagues in their difficult work. At the same time, your response should not denigrate Herbie's current credentials. You might, for example, stress the rigorous academic challenges Herbie would face at PU if he were not prepared to tackle them. Are there alternative pathways you could recommend for Herbie to pursue for future entrance to Prestige University? If so, briefly describe the alternatives in your letter and reference a contact person in the admissions or advising office who could provide more guidance to Herbie and his parents (and be sure to cc this person).

Note. To see how two students responded to Application 6-H, see page 155.

APPLICATION 6-I

Write a Rejection Letter

The background. Writing rejections to job applicants, nonprofits that have submitted grant proposals, and other petitioners for various opportunities is not an easy business. You want to write rejections with tact, both to honor the sensitivities of the recipient of the bad news and to cultivate goodwill among community members whose interest or help you might want to cultivate in the future. (For examples of tactful and harsh college rejection letters in particular, enter "Wall Street Journal" and "college rejection letters" into a search engine.)

In this application, you are an editorial assistant at StarShip Publications, and the editor-in-chief does not have time to write all the rejection letters to the authors whose submissions he and others have reviewed. He has asked you to write a graceful rejection letter to Rachel Adams, who submitted a novel based on her experiences as a competitive surfer.

The purpose. Your supervisor wants the letters from StarShip Publications to be respectful toward the authors who have submitted their work; to suggest plausible, noninsulting reasons for the rejection; and to encourage authors to submit their future projects. This purpose coincides with the company's broader efforts to present a positive professional profile to the many writers struggling to publish their work.

The audience. Creative people are often very sensitive to criticism of any kind. While no recipient will enjoy your rejection letter, make it friendly and respectful, even encouraging if possible. Remember that most aspiring authors, even those who eventually become famous, receive countless rejection letters. Make yours stand out for its sensitivity.

The communication strategy. Begin by expressing the publisher's gratitude that the author submitted her work for consideration. Frame the rejection in a way that does not denigrate the writer's talent. Also, offer a plausible reason for the rejection. For example, Rachel Adams's surfing narrative might not fit with the market demographic for your firm's publications, or StarShip might recently have published a novel with similar content. Conclude with a bit of encouragement regarding Ms. Adams's future projects.

APPLICATION 6-J

Invite a Distinguished Guest to a Campus Event

The background. Businesses often invite prominent people to participate in activities that promote the organization's interests: fund-raising events, product promotions, groundbreaking ceremonies, professional conferences, and so on. A major challenge is that the invitee may not be rewarded monetarily, or may be given only a small honorarium. And if the invited person is rich, famous, or busy enough, even substantial compensation might not be enough to get him or her to participate in your event. Whatever the case, you need to be very persuasive to convince high-profile people to contribute their time and energy.

In this application, you are charged with inviting a local distinguished person to participate in a campus event for your team, group, club, fraternity, sorority, or some other organization of your choice. You would like this person either to join a discussion on a particular topic or to formally present his or her views on an issue of interest to your group. (You can make up an appropriate group if you don't actually belong to one, but be sure to select an actual person as your invitee—perhaps someone of renown.) Assume that the invitee is busy with other obligations and probably receives many similar requests to contribute time and expertise. Your group has $500 available for an honorarium or for other activities to support the planned event (for example, a lunch or dinner meeting with the invited guest).

The purpose. Your goal is to gain the interest of the invited person, to help the invitee feel a connection to your group's interests and purposes, and to feel that she or he has the right background to make a useful contribution.

The audience. Most people you will want to invite are very busy, often in demand for similar groups and events. Sure, you need to flatter the person invited, but primarily you need to arouse his or her interest in your group's purposes. The invitee needs to feel prepared to contribute the requested information and ideas. You also need to show scheduling flexibility to accommodate the guest's busy calendar.

The communication strategy. Use a respectful tone throughout the communication, a tone that reflects your admiration for the invited person and your appreciation for his or her considering your request. Also, make sure to do the following:

- Identify the nature and interests of the group you are representing, giving the invited person enough information to decide whether he or she has the right background to meet your needs.

- Give examples of any related activities your group has sponsored.

- Let the invitee know exactly when and where the event will occur, possibly offering alternative dates.

- Tell the invitee whether there will be other presenters; how long she or he should speak, respond to questions, or both; and who will be in the audience (undergraduate students? faculty? the general public?).

- Stipulate the honorarium if you intend to use some or all of your group's $500 for this purpose.

- Close with a thank-you and a date by which you need a reply.

- Provide your contact information and express your readiness to discuss any of the details.

Note. To see how one student responded to Application 6-J, see page 157.

APPLICATION 6-K

Disinvite Participants to a Focus Group

The background. Sometimes a company will get into an embarrassing situation, despite good intentions and competent planning and management. Issuing explanations and apologies requires very tactful communication. On the one hand, the organization does not want to expose too many details about what went wrong (or just who messed up, if that is the case); on the other hand, the communication needs to explain the misstep and apologize to those affected by it.

In this application, you must figure out how to strike that kind of balance. Here are the details: as a marketing assistant for Full Throttle Research, you were charged with enlisting a focus group of 20 people to assess a new series of television commercials for one of your firm's clients. As it turns out, everyone you invited agreed to participate; further, another assistant also invited a number of participants, and you now have more people than you can accommodate. To make matters worse, you promised all those invited a six-month supply of your client's product, a line of gourmet cookies. You need to decide what strategies to use to rescind some of the invitations.

The purpose. Your goal is to gracefully retract some of the invitations, to briefly explain and apologize for the problem, and to retain the recipients' interest in future focus groups.

The audience. People don't like being disinvited. They feel that they've been snubbed, that you have wasted their time, and that you are not very competent. However, the disinvited people will probably want to participate in future focus groups dealing with television commercials, since they are attracted to being part of media-related events. The promise of an invitation in the future will probably help appease them—as will the complimentary cookies.

The communication strategy. You need to acknowledge the problem without making your company look foolish (and without pointing the finger at any particular staff member). You also want to retain the interest of the disinvited. You must decide for yourself whether to provide the cookie compensations to the disinvited or to withdraw that offer as well.

APPLICATION 6-L

Reassure a Nervous Customer

The background. From a business and legal perspective, providing formal reassurance regarding health and safety matters requires a difficult balance. It's not possible to promise that nothing will ever go wrong. However, you don't want to fuel the anxieties of customers, employees, or other constituencies or just let such concerns go unanswered. The challenge is to foreground the likelihood that things will go as you have planned and prepared for, but to acknowledge that unanticipated things sometimes do occur and that individuals have to exercise care and personal responsibility.

In this application, you are a public-relations rep for the Justin Bieber concert series and have received an inquiry from a nervous ticket buyer. She has purchased a large block of tickets for her daughter's thirteenth birthday and wonders if all the girls will be safe at a massive and energetic event such as this. In your response, you need to provide reasonable reassurance to the customer while also suggesting that the daughter and her friends must take some responsibility for their own safety. Here is the e-mail in which the mother expresses her concerns:

Dear Concert Sponsors:

I am nervous about the upcoming Justin Bieber concert in Missoula, for which I have purchased a block of 17 tickets. I know there have been some crowd-control problems with past Bieber events, and I wonder if I can be confident about this concert on February 27. It's my daughter's 13th birthday, and she and 16 of her friends are really counting on seeing their favorite celebrity. Thanks for your help.

Cindy Jones
Anxious Mom in Missoula

The purpose and audience. See if you can provide some reassurance to the "anxious mom" about the upcoming concert in Missoula—specifying, for instance, the health and safety precautions that will be taken at the concert. At the same time, be careful about making promises about safety that the concert sponsors cannot guarantee.

The communication strategy. Think about your own concert experiences. What is typically provided regarding crowd control, security, medical resources, and so on? Can you sketch these precautions for the nervous mother and at the same time suggest precautions that her daughter and her daughter's friends should take for their own safety and well-being? Again, offer reassurance but no guarantees.

APPLICATION 6-M

Buy Time in a Tricky Situation

The background. Sometimes an oganization needs to buy time in order to gather information, seek further advice, or do some strategic planning—before complying with a request or demand. The interim communication with the person or group making the request should not sound defensive or obstructive but rather provide reasons for needing further time in order to respond more fully.

This application will ask you to tackle such a challenge. In it, you are a paralegal at Pimlico Corporation, and you have received a letter from an attorney, Megan Suya, who is making an accusation and a demand. The attorney for your firm, Wayne Judge, will be unavailable for at least a week, and you are the best-qualified person to write an initial response to Suya. Here is her letter to your company:

Pimlico Corporation
1276 Main Street, Suite C
Long Island, NY 11545

Dear Management of Pimlico Corporation:

My client, Thrombosis Inc., is concerned over your recent decision to curtail use of Thrombosis Inc. as your exclusive catering service. In doing so, you have violated our contractual understanding in effect for the last five years. We request that you immediately turn over all documents related to that understanding, including all e-mails related to this arrangement, all phone-call records for the five-year period, all financial records related to Thrombosis, and all other paper and digitized records that may pertain to the arrangements over the past five years between Pimlico Corp. and Thrombosis Inc. We expect to have all such records delivered to us no later than March 18, 2016.

Sincerely,

Megan Suya, Esq.
AAA Law Firm
1200 Parkhurst Street, Suite D
Albany, NY 12205

The purpose. Your task is to buy some time until the company attorney, Wayne Judge, can look over key documents, consult with some of the Pimlico managers, and decide whether, or to what extent, to comply with Megan Suya's request. Without saying merely that your attorney is unavailable, construct an effective delaying action that does not sound uncooperative or defensive.

The audience. You are writing back to a no-nonsense attorney, so you don't need to labor over a warm-and-fuzzy communication. Just be respectful and cooperative while you briefly describe your company's need for more time. Keep in mind that your letter might be scrutinized by a broader audience in a court, along with other documents in a lawsuit.

The communication strategy. Use a tone and strategy that sound cooperative but that make no promises. Who knows at this point what actual contract or informal agreement might have been in effect for the past five years? You can be sure that your attorney will not want to gather and provide the masses of company documents requested by Suya in the absence of a subpoena, but he might decide to negotiate an agreement with Thrombosis Inc. if it looks as though a lawsuit is looming. In any case, convince Megan Suya that your firm will respectfully consider her concerns and request. Your goal is to buy time without seeming to stonewall Suya and without promising her any particular outcome.

APPLICATION 6-N

Request Permission from an External Constituency

The background. Businesses sometimes request permission to use a privately owned facility or a patented or copyrighted resource. The owner of the resource might be willing to allow limited use free of charge, but more often the owner will want compensation for sharing the resource. The art of requesting permission effectively is a good skill to develop.

In this application, you are helping develop a Web site for your firm Best Practices Inc. The company provides leadership-training services to new or established businesses, helping their staff develop management skills without having to engage in a more formal program, such as an MBA. You would like to use three *Dilbert* cartoons on the company Web site that humorously touch on some of the issues addressed in your training program. How will you go about describing your use of the *Dilbert* images and request a price quotation? (For research, go to the *Dilbert* Web site, pick out three cartoons that deal with training issues, and note how the licensing process works. You can find this information at thedilbertstore.com/pages/about_licensing.)

The purpose. You will need to persuade the licensing firm for the cartoonist, Scott Adams, that your company will make good use of the *Dilbert* material—that is, you will not use the cartoons in some inappropriate context that the *Dilbert* audience would not appreciate (for example, DC Comics might not want to see the copyrighted *Superman* images used to sell toothpaste on television). For more information, see "Restrictions on Customer's Use" on the *Dilbert* licensing Web page.

The audience. Copyright holders are usually very protective of their properties. You need to convince the *Dilbert* group that your use of the cartoons would be in keeping with the spirit of the cartoon series.

The communication strategy. In your letter to the *Dilbert* agents, you will need to define the nature of your company and describe how you propose to use the three cartoons on your Web site. You should also ask for a price quotation.

Student Responses to Selected Applications

Following are sample responses to three of the previous applications: 6-E, 6-H, and 6-J. (Note that the responses to 6-E were written by the author, not by his students.)

Responses to Application 6-E: Resolve a Complaint about Customer Service

Here are two possible responses to the complaint from Charles Jameson (see page 142). The first response (a) is brief and generic and doesn't address the realities of the Rambo situation. Jameson will know that no one is paying attention to his particular interests:

(a)

Mr. Charles Jameson, President
Rambo Corporation
Dearborn, Michigan

Dear Mr. Jameson:

We appreciate your concerns and will give them all the attention they deserve. We will hope to have the issues resolved soon and will contact you when the new equipment is ready to ship. We appreciate your patience.

Sincerely,

Jay Generic
Fescue Ltd.

The second response (b) is much more nuanced and more likely to satisfy this important customer (after all, you do want to retain his business, even if your company was not at fault). Notice that this response is careful not to cast blame on Jameson's employees; it stays above the fray.

(b)

Mr. Charles Jameson, President
Rambo Corporation
Dearborn, Michigan

Dear Mr. Jameson:

I am responding to your e-mail of December 23, 2016, in which you expressed concerns over our last shipment to Rambo Corporation. If you talk further with your Receiving Department, I believe you will find that the stamping machines arrived in good condition, but that the details of final setup and calibration were somewhat confusing. Fescue immediately dispatched two engineers to Rambo Corporation to aid in the setup process, and within a few hours the initial confusions were resolved.

Let me assure you that we will always work with you to resolve any problems that might arise, whatever their cause. Some of our latest technologies are quite complicated, and we are eager to provide the support needed. You have been a good customer with us for seven years, and we hope you will agree that the current frustrations were addressed quickly and professionally. We will do our very best to prevent mishaps or misunderstandings in the future.

If you have any further questions or concerns, please don't hesitate to contact me directly by e-mail at jcares@aol.com or by phone at 882-678-0983. We want to earn your continued trust in our company.

Sincerely,

Jay Cares

Responses to Application 6-H: Respond to a Request from a Privileged Alum

Here are two student responses to Mr. Nudgely, the alum who requested that his nephew be admitted to Prestige University (see page 146). Notice that respondent (a) begins and concludes the letter with thanks to Mr. Nudgely and his family's legacy. She then gracefully offers an academic solution, even acknowledging the promising signs in Herbie's work with Habitat for Humanity. In the final paragraph, she offers a special PU contact person to help Herbie plan his next steps. Overall, the writer has upheld the university's admissions standards, supported the beleaguered director of admissions, and given some insider guidance for Herbie (who will still need to meet acceptable admissions standards).

(a)

Dick Nudgely, CEO
Grandiose Development Corp.
Newton, Massachusetts

Dear Mr. Nudgely:

I'd like to sincerely thank you for your long-term support of Prestige University. We are incredibly lucky to have families like yours affiliated with our university, and your loyalty to us is much appreciated and has not gone unnoticed.

Unfortunately, I will not be able to grant your nephew, Herbie Jones, admission to PU this year. However, I can offer a solution that I hope you find both fair and reasonable. If Herbie attends another school—be it a university or a community college—for one full year and is able to show us academic progress, we'd be more than happy to review Herbie's new application at the end of the year. His year spent with Habitat for Humanity is recognized as showing both maturity and dedication for a selfless cause, and it is looked upon highly by the university. Prestige University would be proud to have a fourth generation of your family attend our school, and I can promise to personally review your nephew's application at the end of next year.

If Herbie would like to discuss these options further with our advising staff, please have him contact Maria Guidestar at 617-987-6543, extension 3. I will forward her a copy of your letter and my proposed solution so that she will be prepared to speak with Herbie. I'd like to thank you again for the time and effort you've put into helping better PU. I look forward to reviewing Herbie's application next year.

Sincerely,

Alyssa Kiandehkian, President
Prestige University
Boston, Massachusetts

The writer of letter (b) on the next page has made similar decisions to those of the writer of letter (a) by framing the response with praise for Mr. Nudgely's contributions and ongoing importance to Prestige University. Remember that Nudgely does not feel like an external constituency; rather, as an alumnus he has made efforts to remain an insider, and PU is *his* alma mater. The writer in (b) decided to say somewhat more about the hard tasks faced by the admissions director, and she asks the director himself to provide information on the appeal process. This second shot at admission might satisfy Mr. Nudgely, or it might instead just push the problem down the road if the appeal is unsuccessful. Always be careful in a complaint resolution not to postpone a likely negative outcome, for this will often increase the complainant's irritation.

(b)

Dick Nudgely
Grandiose Development Corp.
Newton, Massachusetts

Dear Mr. Nudgely,

I would like to thank your family for your continued support of and loyalty to Prestige University. Your wife's and your contributions while serving on the Parents' Council have been invaluable to the continuation of Prestige University's outstanding reputation within the academic community.

I understand your concerns about your nephew Herbie's admission status. I can assure you that our admissions director, Jim Stall, has been inundated with correspondence in the weeks following decision letters being mailed out, and any delay in addressing your concerns is not a reflection on our institution's high regard for your family but simply a result of the stressful and extremely busy admissions season.

Herbie is a fine young man whose talents and accomplishments would surely be an asset to our university. I will personally ensure that Mr. Stall forwards Herbie detailed information on our application appeal process as soon as possible. I appreciate your understanding of the fact that because the responsibilities and decision-making power are divided among the different directors of Prestige, I do not have the authority to override Mr. Stall's decision.

Once again, Prestige University greatly values your family's long-standing support of our institution and its distinguished academic standards. We hope that you will continue to believe in Prestige and contribute to our efforts to expand the opportunities offered here.

Sincerely,

Andrea Michaelian, President
Prestige University
Boston, Massachusetts

Response to Application 6-J: Invite a Distinguished Guest to a Campus Event

Because the following invitation covers all the details outlined in Application 6-J (see page 148), the invited person will know whether he has the right background for the requested topic, wants to address the group identified, has time to prepare for a formal presentation, and can fit the event into his schedule. He should also be pleased by the complimentary, persuasive tone of the invitation.

William M. Daugherty, Chief Investment Officer
WMD Asset Management, LLC
559 San Ysidro Road, Suite I
Santa Barbara, CA 93108

Dear Mr. Daugherty:

My name is Greg Leyrer, and I am a member of the Santa Barbara Finance Connection at UC Santa Barbara. Our club hosts various events aimed at teaching students about the different verticals of finance and has had moderate success throughout the years with helping students break into the industry. With your many years of managing funds and your breadth of experience regarding the securities market, we believe that you could be of great assistance to us and our organization.

We would like to formally invite you to speak at one of our upcoming seminars for around 30 minutes, then to respond to questions. The event will center on opportunities in asset management, specifically with a focus on institutional investing. With your background in finance, you would be the perfect candidate for the job. We understand that you are a very busy individual; we can offer multiple dates and times in order to accommodate what would work best for your schedule. We are hoping to hold our event some weekday evening during the month of February or March. It would mean a lot to our group if you shared your knowledge of the industry by telling us about your own career experiences. We expect around 30 undergraduate Economics and Accounting students to attend the event in South Hall 1432 on the UCSB campus.

Our event will not be complete without a great speaker, and we believe that you are the best person for this task. We would like to offer you a $300 honorarium for your participation. After the event, if your time permits, we would additionally like to take you out to dinner as a token of our appreciation. If you are interested, we would like to hear back from you in order to coordinate the specifics of this seminar. My contact information is provided below. Thank you for your time.

Sincerely,

Greg Leyrer
Member, Santa Barbara Finance Connection
www.sbfinanceconnection.com
gleyrer@umail.ucsb.edu
(925) 324-9673

CHAPTER 7

More Complex Business Writing Projects

Understanding the Challenges of More Complex Writing Projects

Most of the scenarios and applications in this book require the writing of shorter communications, usually of one to two pages. This chapter, however, offers more ambitious writing projects for you to explore, either individually or as part of a team.

Identifying Key Considerations of Complex Projects

The principles and considerations that shape shorter business communications apply to the longer projects as well. You need to determine your purpose(s) in writing, the nature of your audience, and the strategies that will shape an effective communication. There are, however, some differences to take into account, as discussed in the following sections.

The Situation or Scenario

Certain business writing situations take more than one or two pages to address. For example, you might be writing a business plan on behalf of a start-up company, and the plan will have to consider the many factors involved in creating and distributing the company's product or service: the types of customers who might be interested in the product or service, competing companies that offer something similar, proposed advertising strategies, and the finances needed to launch the new company.

The Audience

In more complex writing projects, you are likely to be addressing a much larger audience or multiple audiences—for example, small and large shareholders if you are helping write a company's annual report; expert reviewers if you are competing against other applicants for funds from a foundation; or friendly, skeptical, and even hostile co-workers if you are presenting new salary and benefits structures for your company. Whenever you address a larger, more varied group of readers, you will need to do a more complex audience analysis so that you can make informed decisions about tone and evidence. (Tips on addressing various audiences appear in the applications under "Responding to Real-World Writing Scenarios," pages 164–89.)

Background Research and Preparation

More complex business writing projects might involve extensive planning on your part and will often require you to do library or online research as well. A report analyzing a particular industry (for example, an analysis of the German automotive industry) would require a great deal of research into industry trends and challenges—the changes occurring in customer demographics, the energy and workforce needs of the industry, environmental and safety regulations affecting the industry, and newer technologies that could have a positive or negative impact on the industry.

Some of the longer projects detailed in the "Responding to Real-World Writing Scenarios" section of this chapter (for example, the grant-proposal application on page 175 and the business-plan application on page 182) will introduce you to business-research tools that you might not have encountered in your other academic work. These longer projects will help you learn much more about the types of evidence available to guide all sorts of business decisions.

Further, as you probably know from writing college research papers, finding credible research materials is just the first challenge; of equal importance is analyzing the evidence and organizing it into a coherent and persuasive picture for readers. In essence, you will want to "tell the story" of a company or business situation effectively over a number of pages. To present your evidence effectively, you might want to work from an outline in which you have mapped out this information in a clear, sensible way.

Document Design

Because the possibility for confusion goes up as the length and complexity of a business document increase, you'll need to think carefully about the overall design. Your document could benefit, for example, from formal section titles with subheadings, the occasional use of bullet points, graphics to illustrate complex data, page numbers, and footnotes. As discussed in Chapter 4, "Business Document Design, Formats, and Conventions," thoughtful attention to the organization, formatting, and design of your document will help readers know exactly where they are in your longer piece and where they will go next.

Previewing Longer Writing Projects

This section of *Business Writing Scenarios* previews the types of longer communications that you will be exploring in more detail—and writing—in the "Responding to Real-World Writing Scenarios" section of this chapter. All these communications involve more research than the smaller business documents you have been working with to this point. In the following preview, the projects have been sequenced from relatively straightforward and shorter pieces to longer pieces requiring more research.

A Business-Travel Reimbursement Policy

Most businesses and professional organizations dedicate a lot of time to constructing policies, procedures, and guidelines. The relatively short project in Application 7-A (page 165) will ask you, individually or as a team, to think through the types of travel expenses that your business would and would not support, the preapprovals that might be required, and the record-keeping needed for a successful travel reimbursement. It's a good opportunity for you to anticipate how to make the right things happen for your company (in this case, employee travel that truly contributes to the company's success) and to prevent any abuse of the travel funding provided.

An Ethics Advisory Memo

In the scenario described in Application 7-B (page 167), the CEO of your company has appointed you (or your team) to analyze certain ethical, practical, financial, and legal challenges facing your company and to provide your written recommendations on the best course of action.

A Request for Proposals

In the individual or team project described in Application 7-C (page 170), you will be alerting the local nonprofit community that your philanthropic foundation has a total of $1 million to contribute during the current year to worthy nonprofit organizations. Thus, your request for proposals (RFP) or call for proposals (CFP) should lay out the types of causes you are prepared to fund and the details of the application (or proposal) process.

You will need to define (and name) your foundation and describe the charitable works you want to support in the coming year. You will also need to guide potential applicants concerning the evidence you need to see in their applications or proposals, the deadlines for submitting proposals, and the review process conducted by your foundation. Constructing an RFP will help you experience some of the thinking that goes into a grant writing process, even without your tackling the much longer research process required for writing an actual grant proposal. (The RFP project could also be completed as a preliminary to the grant-proposal project described on page 175.)

A Letter of Inquiry Preceding a Full Grant Proposal

The letter of inquiry, like the RFP, can be completed as its own task or the RFP and the letter can be combined as parts of the full grant writing process. (See Application 7-D, page 171.) It has become customary for philanthropic foundations to request preliminary letters from grant seekers before a full proposal is submitted. This strategy gives both the grant seeker and the foundation a chance to see whether the grant seeker's interests are congruent with those of the foundation. The letter is, in effect, a mini-proposal.

A Grant Proposal

Compared to the RFP project above, the grant-proposal project puts you (or your team) at the other end of the nonprofit funding process. (See Application 7-E, page 175.) You are now a specific nonprofit organization *responding* to an RFP, requesting a certain amount of funding from a foundation in order to support a worthwhile endeavor. This writing project, best undertaken as a team, requires a lot of imagination, much critical thinking, and a good deal of research to make a strong case for your grant request. Your ability to marshal several varieties of evidence and your ability to deploy that evidence persuasively will be put to the test. An especially enjoyable aspect of this project is that it gives you the opportunity to define what good you might be able to do in the world if you had the necessary resources—for example, you might want to achieve environmental goals, improve human health, provide education and training for a particular group, support the arts, or perhaps enhance animal welfare.

A Business Plan

Because it requires extensive research and creative input, the business-plan project is especially suitable for team collaboration. (See Application 7-F, page 182.) The goal is for team members, as entrepreneurs developing a start-up company, to invent a new or substantially improved product or service and to make a well-researched case to potential investors. The gathering of evidence about start-up costs and potential profits, about customer or client demographics, and about any competing companies already in the market makes this an ambitious project.

You will need to think well outside your daily experience to imagine a product or service that is not just a repeat of what's already on the market. Again, the goal is to attract investors to the proposed start-up company through a creative, feasible, and evidence-based plan. This project might conclude with each student team making a PowerPoint or Prezi presentation to classmates, who would serve as an audience of potential investors.

Being an Effective Part of a Team

As already suggested, some of the longer writing projects discussed in the previous section, and explored in more detail on pages 164–89, might be completed most effectively through teamwork. Also, you are highly likely during your work life to explore issues, to reach decisions, and to produce documents through committees, task forces, work groups, and project teams. So it's worth pausing now to consider the rewards and challenges of working collaboratively.

In a college course or in the workforce, teamwork can be as challenging (even frustrating) as it is rewarding. For example, the dynamics of teamwork require team members to set aside or to moderate some of the expectations and behavior patterns that have shaped their work as individuals. Here are a few of the factors you need to keep in mind:

- If you procrastinate in completing your assigned task on a group project, you will have a negative impact on the group enterprise. Thus, your work with a team cannot be self-paced but must honor the interdependence of all the project's contributors. Each team member needs to honor the deadlines established by the team as a whole.

- You must accept that every member of the team will bring different talents, deficits, and levels of interest and commitment to the project. No one individual will have solid strengths in every category. Therefore, it is important to identify, and to make the best use of, each person's strengths and to ameliorate or work around his or her weaknesses.

- No one team member should dominate the team process to the exclusion of others' ideas and responsibilities. The team dynamic will rapidly deteriorate under such circumstances, and any documents produced through the collaboration will likely be less thorough, accurate, and persuasive than is ideal.

- The team cannot allow some members to coast on the energy and contributions of the other members. Your effort should be to draw each team member into the process—whenever possible with positive encouragement and sometimes with offers of help.

- A clear leadership structure is important to every team's success. It's very rare to see a completely egalitarian model operate successfully. Designate a leader or co-leaders to guide the process.

- Individual leaders and co-leaders must set their egos aside. Leaders must be much more interested in the project's goals and results than in receiving personal accolades.

Successful teamwork does not come easily in an undergraduate course, let alone in one's professional life. In both arenas, some members might not meet their responsibilities in a timely way or might not produce work of the best quality. But keep this positive perspective in mind: *Working as a team brings together a group of smart and diversely talented people to pursue a common goal. This network of brainpower and of multiple perspectives is capable of producing wonderfully enriched results.* That's a central reason why so many businesses have embraced the creative potential of teams and task forces.

Responding to Real-World Writing Scenarios

This section will expose you to a range of business writing scenarios, each of which will require you to apply what you've learned about one of the complex documents described on pages 161–63. Recognizing the special challenges of such communications, we provide guidance not only on purpose, audience, and communication strategy but also on the particular features of the various documents.

Applying What You've Learned

The following six applications ask you as individuals or in teams to think and research your way through some complex challenges.

APPLICATION 7-A

Draft a Business-Travel Reimbursement Policy

The background. Professional organizations generate an abundance of policies and guidelines intended to define company procedures, ensure equitable treatment of employees, and minimize the organization's legal vulnerabilities. Policies may specify criteria for advancement or termination, health and safety standards in the workplace, consequences of sexual harassment or bullying, rules regarding the use of sick leave and vacation days, benefits and salary levels, manufacturing and service standards, and so on.

The most productive approach to creating or clarifying policies is to regard them as a positive means to achieve an organization's goals, not just a list of restrictions. So, first define the goals to be attained, and then write policies and procedures that are likely to achieve those ends and, at the same time, prevent inequities, liabilities, and the misuse of resources. Thus, a good policy is both positive in its goal orientation and also protective of both the employees' and the company's interests.

In this application, you are assistant to the director of human resources and are helping compose a policy statement, accompanied by procedural guidelines, on business-related travel expenses for the company X-Factor. The policy statement and guidelines should be one-and-a-half to three single-spaced pages. You will also draft an expense reimbursement form that implements certain aspects of the policy and guidelines.

X-Factor's top management wants to tighten and clarify business-travel policies because of cost concerns. In the past five years, the company's travel expenses have tripled, to an annual total of $33,000 (for a company with 55 managerial staff). The process for approving travel on behalf of X-Factor has been quite loose, and seldom have any travel expenses been questioned or denied.

The purpose. X-Factor is seeking better cost control but wants also to support managers' legitimate needs to attend professional and trade conferences, to meet with major clients, and to carry out other important business. Crafting a business-travel policy that is both fair and financially prudent will require asking the following questions:

- What types of travel are truly essential to the company's success?

- What preapproval structure should be in place to ensure that travel funds will be spent for the best professional purposes?

- What per diem or other expenditure guidelines or restrictions would be useful for keeping travel expenses under control?

- What process will be followed for submitting travel-expense vouchers and receiving reimbursement?

The audience. Managers at X-Factor have enjoyed the liberal travel policies of recent years and will need to be shown the sensible financial reasons for tightening the approval and reimbursement policies. They will also appreciate a clear statement regarding the types of business travel that are likely to be approved and the company goals to be served by business travel.

The communication strategy. Business-related policies are essentially efforts to promote a company's goals and interests and to prevent misunderstandings and problems (such as misuse of resources, health and safety problems, legal vulnerabilities, and salary and benefits inequities within employee categories). The top management at X-Factor wants to preserve the excellent morale among its managers as it announces a tighter policy on travel expenses. The tone of your draft should reflect the respect and trust that prevails at X-Factor. So address the audience as colleagues, as genuine partners in the success of X-Factor. Provide them with the evidence that will help them understand the need for a clearer policy, and avoid making the preapproval processes too complex and burdensome.

The parts of the policy document. The policy document should consist of the following:

- A descriptive title of the policy

- Background considerations: Describe the challenges, problems, and confusions that the policy addresses. In essence, you are answering this core question: why do we need the policy?

- The policy: As much as possible, present the policy in positive terms, defining the good or useful goals it will achieve. Also, clarify the spending limits for transportation, hotels, meals, conference fees, and other expenses, and explain why these limits are necessary.

- Procedural guidelines: Outline the steps for implementing the policy in an orderly, timely, and equitable fashion. Articulate the criteria to be used for reviewing travel requests, the advance notice needed for initial approvals, and the deadlines for seeking reimbursement. Also, identify the supervisor(s) who will review travel and reimbursement requests.

The expense reimbursement form. This should be a one-page (or shorter) form for employees to submit for manager review after their travel is completed. It will need to be accompanied by expense receipts.

You will need to design a foolproof form, one that is easy to fill out and that makes reimbursement requests conform with the travel policy and guidelines. (For resources that will help you create such a form, see the next section.)

Further help on this project. You can seek out further online resources that will help you understand the procedural and financial issues associated with business-travel policies. They might examine such considerations as who may travel and for what purposes, what preapprovals are required for travel, what are the expense limits for travel reimbursements, what expense records are required for reimbursement, and how employees submit reimbursement requests. Note that the samples and guidelines you find online may be much longer and more complex than the policy you are asked to create in this application. Also note that you should not copy any of these sample online materials; rather, use them to stimulate your own thinking and sense of document design. Conduct a Web search using the following search terms to find additional resources.

- Travel Policies and Forms at the University of Illinois
- Casto's Sample Corporate Travel Policy
- Search for "Business/company/corporate travel policy" to find many more examples

APPLICATION 7-B
Write an Ethics Advisory Memo

The background. Ethics questions arise so often in professional settings that every month brings news of the latest ethical lapse. Typically, the central point of tension is between honesty and fairness on the one hand, and profit or personal motives on the other. Undoubtedly, we never hear about the many more routine, ethical commitments that good businesses make every day to their employees, customers, and stockholders.

This application will ask you to think through an ethical quandary at a company and suggest solutions that will somehow uphold honorable standards without imperiling the company's longer-term viability. Specifically, Meg Whitely, the CEO of Kanga Toy Company, has appointed you and several other staff at Kanga to advise her—in the form of a memo—on a rapidly emerging ethical and financial dilemma.

Kanga has been a leading U.S. producer of safe and creative toys for more than 30 years, especially targeting children between the ages of one and seven years. In recent years, however, sales of the company's traditional lines of toys have declined dramatically.

A positive development has been Kanga's recent launch of LectroKitty, which has brought new financial life to the company. The fast-selling

LectroKitty is a cute robotic cat that meows when the toy's child-owner approaches, periodically sharpens its claws on a scratching pad, ingests plastic kibbles from a food bowl at programmed times in the day, leaps after cat toys when given certain commands, and purrs and bats its eyes when petted by the child. All of the parts needed for these automated responses are supplied by Kanga Toy Company.

During the past few months, however, some customers have raised concerns about LectroKitty. A dozen parents posted comments to the company's Facebook page complaining that their children had swallowed and choked on the plastic kibbles. Several other parents expressed concern that the toy gave a substantial electric shock to their child when the child kissed LectroKitty on its head (the toy operates on a miniature 16-volt lithium ion battery developed abroad especially for Kanga Toys).

Sales thus far of the LectroKitty have been 350,000 units, for a gross sales revenue of more than $12 million. Thus, the volume of customer complaints has been very small in proportion to the sales volume. In total, two children experienced burns from the electric shock serious enough to warrant a trip to the hospital, and two children were taken to the emergency room after choking on the plastic kibbles.

Given the extraordinary sales success of LectroKitty, Kanga Toy Company is designing other robotic toys to replace its older product line. In just the past month, three of the largest retailers in the United States have expressed interest in distributing LectroKitty in their stores.

Some unhappy parents communicated with a major consumer watchdog group and with the U.S. Consumer Product Safety Commission (CPSC), whose mission is to ensure the safety of consumer products. Yesterday the CPSC phoned CEO Whitely to make a preliminary inquiry about the reported problems with LectroKitty. Ms. Whitely was not available to speak with the CPSC, but Whitely did speak with her chief design engineer, Suzanne Ciao, who believes that she and her team can analyze and fix the LectroKitty problems if given around three months. The attorney for Kanga Toy Company, Mike Seidler, has expressed serious concerns over the damage to Kanga Toys' reputation if further problems are reported, if Kanga is sued by angry parents, or if the CPSC orders a recall of LectroKitty.

Yet another wrinkle in this scenario is the profit-sharing promise made to Kanga's 230 unionized employees when they last negotiated their contract. Because the employees had taken salary reductions for two years when Kanga profits were declining, the Kanga management formally agreed to share profits that might be derived from the sales of LectroKitty.

The purpose. In your memo, you and the other appointed members of Meg Whitely's management team need to advise her on the best course of action to take. You will need to explore complex issues of safety, ethics, company reputation, and legal liabilities in relation to the company's financial well-being and the contractual interests of its employees. Based on this analysis, you will then make written recommendations to the CEO, as she requested.

The audience. Meg Whitely, the CEO, is the immediate audience for your memo, and she will undoubtedly share your recommendations with other key members of the management staff. She will care about the logic of your analysis, your ability to balance the several contending interests in this situation, and the clarity and cogency of your recommendations. In other words, she will want to see your reasoning process in addition to your conclusions.

The communication strategy. You are writing to a knowledgeable insider, the CEO; thus, you will need only a very brief summary of the LectroKitty situation. Of more interest to the CEO will be your understanding of what's at stake for the company and the insights and arguments you provide as you move toward your final recommendations. Be sure that you *do* make expedient recommendations to the CEO; don't just conclude that the issues require further study. Whitely needs very quickly to decide what to tell the CPSC and others who might inquire about the safety of LectroKitty, how to respond to the major retailers who want to distribute the toy, whether to stop shipping LectroKitty until the problem is analyzed and fixed, whether to recall the product with a general public announcement, and what to say to the employees who have been counting on the profit-sharing agreement.

Contents of the ethics advisory memo. Rather than write a continuous narrative in your memo, you should create some appropriate subsection titles to make the document easier to read. For example, you might divide the memo into sections like these:

- "Background" (in which you briefly describe your group's assignment from the CEO and sketch the central challenges faced by the company)

- "Ethical Considerations" (in which you analyze the central philosophical or ethical and practical concerns raised by the LectroKitty situation)

- "Conclusions and Recommendations" (in which you provide the CEO with the results of your examination and recommend a course of action)

These are just suggestions for a memo structure. You may develop your own structure and titles as the team process warrants. But be sure to keep your CEO audience in mind: you want her to perceive quickly the logic, cogency, and utility of your advisory report.

Further help on this project. The ethics advisory memo project will be greatly enriched if you view, and discuss in class, the online video *Ethics in America II: Risk, Reward, and Responsibility: Ethics in Business*, produced by the Fred Friendly Seminars with the support of the Columbia University School of Journalism. (For more information, visit www .learner.org and enter "Ethics in America II" in the search field.)

The video offers several hypothetical business-ethics scenarios for discussion by a panel of experts in business leadership, ethics, politics, economics, and law. You will notice that these experienced people often disagree about the best ethical and business responses to the scenarios presented to them. There is often no easy solution that balances all of the competing claims of ethics and business pragmatics. The lively debates recorded in the video will, however, shed light on the Kanga Toy Company scenario and on real-world ethical situations you are likely to face in your professional careers.

Also available at www.learner.org is an excellent study guide for the series *Ethics in America II: The Ethics in America Study Guide* by Dr. Lisa H. Newton, a professor of philosophy at Fairfield University.

At the end of this chapter, you will find one of Newton's business-ethics scenarios and a sample advisory memo that students wrote in response to it (see pages 189–94).

APPLICATION 7-C

Write a Request for Proposals (RFP)

The background. In this application, you are the grants coordinator for a philanthropic foundation, and you need to write the foundation's annual RFP. An RFP lets nonprofit organizations know that you have funds available for certain worthy causes and are ready to receive proposals for your foundation's review. This application can be undertaken individually or as a team.

The purpose. You need to answer this basic question: "If I represented a foundation with $1 million to contribute this coming year to worthwhile nonprofit organizations, what sorts of good works would I want to underwrite?" Once you've answered this question, draft an RFP that defines the following:

- the types of charitable enterprises that your foundation wants to support (for example, projects supporting the arts, education, the environment, or human health and well-being)
- the typical size of the grants
- eligibility requirements for the nonprofit organizations or individuals making proposals

- the content expected in the proposals
- the application and review process and the deadlines for submitting materials to your foundation

The audience. Many individuals and organizations are looking for funding, so you will need to be very explicit about the activities you want to support. Your RFP should be encouraging and respectful, but it should also be very clear about the eligibility criteria you will use in the review process, the size of the grants to be awarded, and the date by which you must receive all proposals and supporting materials. You don't want to be flooded with money seekers who don't understand your foundation's expectations.

The communication strategy. The RFP's statements about the purposes and review criteria for your foundation's grants should be brief and clear, allowing for no ambiguity. For example, if you state vaguely that the grants seek to "support activities that enrich local communities," you will be swamped with a huge array of applications, many of which will not fit your actual intentions.

You want to be certain that the grant proposals you receive will be a good fit with your organization's philanthropic and civic interests and that all the information needed for the review process will be provided. You also want to see evidence that the applicants have a clear and feasible plan for spending your funds effectively, so be sure to build such accountability expectations into the RFP. For example, ask the applicants for an explicit budget and plan of action, and ask how they will provide evidence (within a defined time frame) that they have spent the grant money appropriately and with a demonstrable impact.

Further help on this project. The Non-Profit Guides Web site, at www .npguides.org, provides good examples of RFPs. Also, a sample of a student-written RFP is included at the end of this chapter (see page 195). Whatever RFP models you review for ideas about format and content, be sure to devise your own nonprofit project. Make the RFP reflect *your* interests in doing something good for the world.

APPLICATION 7-D

Write a Letter of Inquiry Preceding a Full Grant Proposal

The background. In this application, you are representing a nonprofit organization seeking a grant to support your organization's charitable goals. Submitting a preliminary letter of inquiry is a very common step in current grant writing processes: the letter allows both the foundation supplying the funding and the grant applicant to see whether there is a reasonable fit between the expectations of both organizations—a good reason to move forward with the more formal process.

Apart from its connection to a grant writing endeavor, your experience in composing a letter of inquiry might help you, in the future, propose your ideas and interests succinctly to a local government entity, to the management team at your company, or to a task force within your company. The letter is an efficient means of testing the waters before you invest a great deal of time in creating a fully developed plan or proposal.

In this application, you are writing—as an individual or as part of a team—a letter of inquiry to the New Horizons Foundation (whose RFP appears on page 173). In this letter, you want to show the foundation that you have a worthwhile and feasible plan for a nonprofit project and that you should be encouraged to submit the full grant proposal by the foundation's deadline.

You will need to imagine or create a worthwhile nonprofit project. You will also need to do some preliminary research on the population that would benefit from the project in order to see what is actually needed and feasible. For example, if you wanted to help the homeless in Youngstown, Ohio, you would need to research the homeless demographic in the Youngstown area, getting answers to questions like these:

- How many homeless are in that area?

- What are their ages and ethnicities?

- What are the reasons for their homelessness, and what are their various needs?

- What is already being done for this population?

Then, to find a good niche for your own nonprofit efforts, research what types of services and interventions seem to work for this homeless population, and what still remains to be done.

Let's assume in this application that your nonprofit organization already has a five-year history of successful projects and that you have raised $28,000 toward the new project that you want to propose to the New Horizons Foundation. Thus, in the letter you will be able to make some (imagined but plausible) claims about your nonprofit endeavors.

First take a close look at the New Horizons Foundation RFP on the next page to see what they are offering to and requiring from applicants.

The purpose. Your letter of inquiry must attend carefully to the New Horizons RFP and make a compelling case so that the foundation will encourage you to submit the full proposal. That's your central purpose. To be convincing, you will need some specific ideas to demonstrate that you have thought closely and creatively about your proposed project and that you have done enough preliminary research to show that the project is needed by a particular constituency and is both feasible and affordable. (You should include footnotes that specify the sources of research data that you cite in your letter. For examples of footnotes, see the sample letter of inquiry on pages 198–202.)

NEW HORIZONS FOUNDATION
2211 Berkshire Road
Pasadena, California 91104

February 22, 2016

Request for Proposals (RFP)

The New Horizons Foundation (NHF) was established in 1995 and has thus far in its history awarded a total of $8.5 million to nonprofit organizations around the United States. The Foundation provides grants to nonprofit organizations in support of education, health and wellness, environmental protection and sustainability, the arts, energy conservation and alternative energy development, aid to the elderly, and other worthwhile community-based projects.

Applications responding to our April 2016 RFP may request up to $100,000 for the proposed project. However, available funding is contingent upon the amount of funding in the NHF budget for fiscal year 2017.

A preliminary *letter of inquiry* from the grant seeker must be submitted to the NHF by March 31, 2016. The letter should describe succinctly the need for the proposed project and the grant seeker's planned response to the defined need, and it should provide a preliminary budget (with explanatory narrative) that would meet the need over a particular period of time.

The NHF will respond to letters of inquiry no later than April 29, 2016, and may, at the discretion of the NHF, invite full grant proposals from some organizations. The full grant applications, with all supporting material, must then be received by the NHF no later than June 30, 2016.

For questions regarding your eligibility for a grant, or for guidance on any stage of the application process, please consult Jayne Carrie, Senior Grant Coordinator for the NHF, at jcarrie@newhorizons.org.

The audience. Philanthropic individuals and foundations that fund non-profit projects want to make good things happen in the world, and they want to lend their financial support to worthy organizations. At the same time, they want to be sure that their money will be spent wisely and effectively—that you, the petitioner, have a clear understanding of the need being addressed, the solution, and the costs. They will expect to see in your letter a solid understanding of the issues you want to address and a feasible and affordable plan of action.

The communication strategy. Combine your creative thinking about the desired nonprofit project with some preliminary research. (You should identify three or four credible information sources that you refer to in your letter, and be sure to provide full information on the sources in your footnotes.) Remember, the central strategy is to combine your idealistic desire to accomplish good with a reality-based assessment of how you would actually make good things happen.

Contents of the letter of inquiry. Using the block letter style format (see pages 88–89), address the letter of inquiry to the contact person referenced in the New Horizons Foundation RFP (Jayne Carrie, Senior Grant Coordinator) and provide the full address of the foundation.

Write a letter of one-and-a-half to two-and-a-half single-spaced pages and include the following components: an introduction, a description of your organization, a statement of need, your methodology for addressing the problems or issues you have identified, a brief discussion of other funding sources, and a final summary. The overarching goal is to "tell the story" of your organization's successes and aspirations.

Here are more details on the parts of the letter:

- **Introduction.** The purpose of the introduction is to provide an executive summary for the letter. It includes the name of your organization, the amount of money needed or requested, and a very brief description of the project. This section should identify why your team is a good fit for the project by mentioning your method of addressing the identified need, your qualifications, and a possible timetable for implementing your project. In some cases, a philanthropic foundation will not issue a specific RFP; in such cases, you will need to research the foundation's funding interests and the procedures they expect you to follow.

- **Description of your organization.** This concise description should provide a very brief history of your organization, describe your current program(s), and demonstrate that your organization is able to meet the stated need. Make clear how you plan to accomplish your new (or expanded) goals with the requested funding.

- **Statement of need.** Why should your project receive funding? This essential element of the letter must answer that question by convincing the foundation both that the need you have identified is important and that your proposal is the best way to address it. The statement of need should identify an issue, justify your proposed solution, and articulate who will benefit from the project. It should include information on the population and geographic area your project will address. You should support your proposal with statistical data and specific examples, where appropriate. For example, a letter of inquiry about a project aimed at helping homeless people in Youngstown, Ohio, might include statistics on the growth in the homeless population in that city. It might also give examples of difficulties that population is facing. This information should clearly be based on your individual (or team) preliminary research and should be documented with footnotes.

- **Methodology.** How will you accomplish your stated goals? Outline your plan for the project in a coherent and organized way, emphasizing what makes your approach stand out as well as what makes it achievable. What specific activities will you undertake to achieve your desired objectives? Who from your organization will be involved? Convince your reader not only that your project is a great idea, but that you have a clear and comprehensive plan for implementing it.

- **Other funding sources.** Make reference here to the $28,000 that, as noted earlier, you have already raised to support your project (and state again the additional amount of funding needed from the New Horizons Foundation). Also include an overview of the total funding needs of the project.

- **Summary.** Close by quickly summarizing your project goals and by offering to provide any additional information needed (provide a contact name, address, and e-mail). Thank your readers for their consideration. Conclude, as you would in any formal letter, with a "sincerely" and your name/signature.

Further help on this project. Grant Space provides a reasonably good sample letter asking the Blue Ridge Foundation to support internships for at-risk inner-city youth. To view this letter, visit grantspace.org and enter "Blue Ridge Foundation" in the search field.

You might also want to refer to the student-generated letter of inquiry that appears at the end of this chapter (see page 198).

APPLICATION 7-E

Write a Grant Proposal

The background. At its core, this project, presented as a team activity, is about obtaining funds to support a worthwhile goal. Many organizations, whether nonprofit or for-profit, raise money through investors and donors. For example, arts organizations usually survive because of their successful grant proposals and donor solicitations; science faculty in research universities engage annually in raising funds to continue their research projects; private businesses turn to investors to support new-product development; and entrepreneurs seek investors for their start-up companies. Your writing a grant proposal will enhance your skills in researching the evidence to support a financial request and in making a compelling case for your professional aspirations.

In this application, your team is a nonprofit organization seeking a grant from the New Horizons Foundation for a worthwhile philanthropic project. (See this foundation's RFP on page 173.) As part of this project, you have been asked to write a preliminary letter of inquiry to the foundation (as described in Application 7-D), and your instructor might also require draft stages for the full grant proposal.

Here are the underlying assumptions and ground rules for this application:

- You have been part of an established nonprofit group for five years and have one or more successful projects or activities that you can briefly describe to the NHF as you make your case.

- You have already raised $28,000 for your current proposal to the NHF. (You should research actual, relevant donor groups to name in connection with this information.)

- You can request up to $100,000 from the NHF, meaning that you could have a total of $128,000 to pursue your philanthropic goals.

- You must be able to show the NHF how you will develop and sustain your proposed project over at least three years after receiving funding.

Please note that in this activity you are writing a hybrid version of an actual grant proposal, one that prepares you for the real-world application process and that also responds to the academic goals of your college or university:

- Some real-world grant proposals can be as brief as 1 page or as long as 100 pages, though the typical length is approximately 5 to 10 pages. The proposal created by your team should be 20 to 30 pages, single-spaced, in order for each team member to make significant research and writing contributions to the project.

- Your grant proposal should be rooted in and supported by research; you are required to use footnote citations and to include a Works Cited (bibliography) section as well. This research expectation might be more rigorous than in some real-world proposals. (For more details on creating a Works Cited section, see page 180.)

- Your grant proposal should provide a narrative explanation of the projected budget that relies less on the numbers themselves and more on your written analysis and explanation of what these numbers demonstrate. (For a good example of this type of narrative, refer to the Budget section we've excerpted from a student grant proposal; see pages 210–13.)

- You must include at least four graphics in your grant proposal to help make your case to the NHF. (For more information, see the suggestions on "Incorporating Visual Materials into Your Text" that appear in Chapter 4, pages 93–96.)

- Pay close attention to creating a document design that makes the proposal visually appealing and easy for readers to follow.

The purpose. Your purpose in writing this grant proposal is twofold: first, to demonstrate your creative thinking and research skills as you invent a worthwhile and feasible nonprofit project; second, to use your reasoning and research capabilities to devise a plan of action (with attendant financial costs) that will persuade the NHF to fund your project.

To stimulate your thinking, here are some possible nonprofit project ideas for you to consider. You may develop one of these ideas or move entirely in your own direction.

Education

- Computers on wheels: an info-technology bus for school neighborhoods (after school)
- Free, after-school, online tutoring for junior-high and high school students offered by college students
- Literacy project for adults (especially linked with jobs and careers)

Energy/environment

- Solar-energy advising for homeowners and small businesses

Animals

- Pet-care assistance and information on pet care for the elderly
- Pet-care or adoption services for pets stranded after storms or other disasters

The arts

- A service that allows individuals to borrow original works of art (and reproductions) for temporary display in their homes

Health

- After-school healthy eating workshops (and nutritious food prep) for junior-high students

Economic concerns

- Job-search counseling for the unemployed and underemployed
- University-based workshops for students regarding personal budgeting, savings and investment, use of credit cards and loans, and other personal-finance issues

International understanding

- Weekly reading and discussion groups concerning the United States and the world

Community service

- Free (or low-cost) handy-person and painting services for the elderly or disabled

The audience. You are writing to people who are committed to philanthropy but who also need to see substantial evidence that you have identified an important need in your community and have devised an effective and affordable plan to meet that need. You will need to persuade the reviewers at the NHF through the factual and anecdotal evidence you provide, through your creative and sensible solution to the issues identified, and through the clarity, coherence, and storytelling power of your writing.

The communication strategy. With any grant proposal, you are competing with a number of other worthy organizations for funding. To distinguish your request from those of other petitioners, you will need to provide research-based evidence and compelling anecdotes that show the need for your nonprofit project, offer creative solutions for addressing this need, and produce a document that is clear and well designed.

Contents of the grant proposal. There is no single format used in all grant proposals, but the sections described below offer an effective approach to the overall structure:

- **Title page (1 page).** Design an attractive and informative title page that includes the purpose of the proposal, the sender (your organization's name), the recipient (the New Horizons Foundation), and the date of submission.

- **Cover letter (1 page).** Sometimes called a "transmittal letter," this document uses a standard letter format, with the recipient's name and title and the foundation address. The cover letter states, succinctly, why your proposal meets the goals of the New Horizons Foundation as announced in its RFP. The letter also briefly defines your nonprofit project and the amount of funding you are requesting. The leader or facilitator of your team should sign the letter and provide contact information for your organization.

- **Table of contents (1 page).** This page should designate the major sections of the proposal and the page on which each section begins. Your goal is to make it easy for readers to navigate through the document.

- **Executive Summary (2 pages).** The Executive Summary provides a succinct overview of your organization's history and successes, the goals of your current nonprofit project, the methods you will use to address the identified problem or issue, and the basics of the budget needed for the project to be successful. The Executive Summary must be focused and engaging, or busy reviewers might not read the rest of your proposal! (For an example of an Executive Summary, see pages 205–7.)

- **Problem or needs statement (5 to 7 pages).** This section must be especially rich in supporting research. With the aid of available data, expert testimony, and anecdotes, define the problem or need you are addressing. What exactly is the problem, need, or shortfall that your project will address? If you are, for example, planning to help a certain group of people, specify such demographic details as their ages and ethnicities, where they live, their interests, their socioeconomic standing, and so on. Don't define the target need or population so broadly that the problem seems insurmountable.

- **Program or project description and methodology (5 to 7 pages).** This section, too, will require a good deal of research, along with a lot of creative and logistical thinking. The creative aspect is your ability to think of ways that might truly remediate the problem you have defined. The logistical thinking shows that you have weighed all the practical steps and methods needed to address the problem successfully. In other words, how would you actually make "X" happen if you had the funds you are requesting?

- **Means of evaluating the project's success (2 to 4 pages).** A grant organization wants to know what measures you will use to demonstrate that your program is actually working—that is, having the desired impact. The foundation wants to see *accountability* to be sure it is getting its money's worth. You should include a projected time line for your reporting of results (perhaps in defined increments at the end of each year for a three-year period).

- **Budget with a narrative explanation, and your plans for future funding (2 to 3 pages).** To construct a realistic budget, don't just guess about costs; rather, do some research on the costs of facilities you will need, of advertising the program, of supplies and services needed, of staff salaries, and so on. You will quickly find that the maximum you might have available for your program ($128,000) won't stretch as far as you would like. If this proves to be the case as you construct the budget, adjust the size and coverage of your program proposal to fit the financial constraints. It's better to devote sufficient resources to a more limited program than to obtain mediocre results with a large, underfunded program. One good strategy is to present your agenda as a pilot program of limited scope; in the future, you would then be able to build on the successes of the pilot effort.

Throughout the budget construction, don't assume that the numerical data will speak for itself. Instead, frame the key financial figures within your own narrative analysis of the numbers. Let your clear prose and careful interpretation show the reviewers what they need to know about your financial projections.

Another important part of the budget section is your projection of future funding. Most philanthropic foundations don't want you to rely forever on their resources, to return to them every year for more money. They hope that you can reach a relatively sustainable state after a few years. Thus, you need to describe your ideas for obtaining project support in the future.

- **Conclusion (1 to 2 pages).** This is your final pitch, your last opportunity to bring the major strands of your proposal together in a compelling manner. You have, in effect, come to the end of the broader story line that links the parts of a persuasive proposal together. Be sure to restate the amount of funding you are requesting and the wonderful impact this resource will have. Your concluding sentences can pump up the rhetoric a bit to achieve a final inspiring moment.

- **Works Cited/Bibliography (2 to 3 pages).** Many real-world business writers give scant attention to acknowledging their sources of information and ideas, but in an educational setting you must be meticulous about giving credit where it's due. That's a core value of scholarship and key to the advancement of learning: relying on the thinking and research of those who have made the journey before you and adding your own contributions as well. Your readiness to make use of, and acknowledge, others' contributions over the years is a sign of your strength as a student, researcher, and writer.

  You have already included full and accurate footnotes for the research that informs your grant proposal. Now you need to construct an alphabetical list of all the resources you have referenced in the document. Use the footnote and bibliography formats requested by your instructor. One excellent guide to footnote and bibliography formats is the Harvard Business School *Citation Guide*. To access this resource, visit www.library.hbs.edu and enter "citation guide" in the search field.

Further help on this project. At the end of this chapter (see page 204), we have included excerpts from a recent grant proposal written by a team of university students. This is probably the most instructive place to start.

Although the sample proposals available online don't quite fit the hybrid model developed in this chapter, you can get some useful tips on the strategies used in professional grant proposals at these sites:

- **Non-Profit Guides: Grant Writing Tools for Non-Profit Organizations**, at www.npguides.org, presents two sample grants and many fine tips for grant writers.

- **GrantSpace.org** posts 15 real-world proposals and parts of proposals. See "Sample Documents" under the site's "Tools" menu.

- **Scot Brannon**, a professional grant writer, provides several of his successful proposals online. He is especially adept at incorporating financial and other numerical data and at using charts and graphs. To see these proposals, visit thegrantdoctor.com, and click on the "Sample Proposals" link.

Your instructor or your college librarians, or both, can introduce you to excellent online resources for researching the content for your grant project. Here are just a few of the best online resources:

- **GuideStar**, at www.guidestar.org, tracks detailed financial and other information on 1.8 million nonprofit organizations. This resource can give you an inside view of how nonprofits are organized and of how they spend their resources.

- **The National Center for Charitable Statistics**, at http://nccs .urban.org/, provides enormous amounts of information on nonprofits comparable in your targeted geographical area that have goals comparable to your own. To access this information, click on this Web site's "Nonprofits" link.

- **American FactFinder**, at factfinder2.census.gov, is a giant warehouse for demographic information that can pinpoint the region in which you want to operate your nonprofit.

- **ProQuest Statistical Insight** is a compendium of all sorts of socioeconomic and cultural statistics that can help you understand the issues affecting the target population of your nonprofit.

- **LexisNexis Academic and Business Source Complete** are probably the most useful of the giant databases for finding up-to-date analyses of the factors that shape the needs and problems addressed by your nonprofit.

Note: Several of the listed databases, including ProQuest Statistical Insight, LexisNexis Academic, and Business Source Complete, can be searched only by means of a paid subscription. Your college library can probably provide access for you.

APPLICATION 7-F

Write a Business Plan for a Start-Up

The background for business plans. A well-conceived business plan is an essential document for convincing sophisticated investors to help you start a new company. A business plan is also an important tool used by established businesses for strategic planning. The most famous business plans of the twenty-first century were those created by General Motors (GM), Chrysler, and the Ford Motor Company in 2008 during the recession. The U.S. Congress required these companies to submit the plans before the Senate Banking Committee would consider providing billions of dollars in so-called bailout funds to the American automobile industry. (The plans were intended to show how the auto companies would run more productive, cost-effective operations.) As widely reported in the press, Chrysler and GM actually "flunked" this exercise and were required to submit improved business plans in 2009. Ford had requested only a line of credit from the Feds, and its plan passed muster (Ford ended up not using the available funds). You can view the 2008 business plans online by conducting a Web search using the terms "General Motors/Ford/Chrysler" and "Business Plan 2008."

The background for this application. The business-plan project you will tackle now is an ambitious undertaking. You will work with other students over the course of a term to create your own start-up company. Your entrepreneurial talents will be put to the test as you imagine and research all aspects of creating a new product or service from scratch. Your research for this business plan will acquaint you with many of the marketing, demographic, financial, production, and management considerations that affect businesses of every sort. And your abilities to work as a team to produce a clear and persuasive document will be critical to your convincing investors to support your enterprise.

This project has been adapted to your focus on writing, research, and reasoned analysis in a university setting. Thus, the report you create will be longer than many actual business plans, will stress the quality of your research and writing, and will examine some issues in greater or less detail than might be the case in a more traditional business plan.

You may never write a formal business plan in your professional life. However, whenever your professional activities involve strategic planning, financial analysis, or the creation of policies and procedures, you will draw on the research and writing skills that this business-plan project is designed to build. You will learn a great deal about the concepts and practicalities that contribute to the success or failure of an actual business, and you will become familiar with professional research tools that will help you explore many business challenges and opportunities in the future.

Now, let's turn to the specifics of this project. Imagine that you and your team members are a group of young professionals who have a great idea for a new product or service. You know you need to do a lot of thinking and research about the exact nature of the product or service, the demographic profile of your potential customers, the competition from companies manufacturing similar products or offering similar services, advertising and marketing strategies, staffing needs and salaries, the facilities (physical or Internet-based, or both) you will need, the means of delivering the product or service, a realistic time line for opening the new business, and so on. You will also need to figure out a realistic projection of costs and revenues (usually over a three-year period) to see whether you will have sufficient revenues to build a successful business—and to attract investors, who expect a good return on their investment (ROI).

Depending on your instructor's decisions and the time available in your term of study, it will take at least four to six weeks to complete all the research and analysis tasks and to produce a business plan of 25 to 35 pages (single-spaced).

The purpose. There are two main purposes in writing the business plan, one for you and one for the potential investors who will read your plan. Your personal goal is to find out, through your creative thinking, research, and writing, what it will actually take to launch your business and make money from it. (In the real world, it's much better to fail on paper than to fail personally and financially with a poorly conceived business venture.) Your second major purpose is to convince potential investors that you not only have a good idea for a business but that you also know how to turn that idea into a profitable reality.

Use your imagination in devising a service or product for your business plan. To stimulate your thinking, here are some recent topics from business plans produced by undergraduate students at the University of California, Santa Barbara. The locations are mentioned because the students' market research revealed an appropriate customer base residing in or near the areas chosen for the start-up company:

- **Gourmet2Go.com** (located in Bellaire, Texas). Gourmet take-out groceries and prepared meals for busy professionals.

- **Cardinal Vending** (located in Columbus, Ohio). School-supplies vending machines located in university residence halls.

- **Skate Unlimited** (located in Phoenix, Arizona). Easy-to-install indoor skateboard parks.

- **Helio Energy Solutions** (located in Glendale, Arizona). Competitive-cost solar installations for small businesses.

- **Freedom Rides** (located in Washington, D.C.). On-demand rental bikes for city transportation.

- **Calendar Couture** (located in New York City). High fashion through a monthly subscription service for women.

- **Sunshine Styles** (located in Punta Gorda, Florida). Contemporary fashions for older women.

- **TrakBack Inc.** (located in North Las Vegas, Nevada). GPS tracking for vehicles and construction equipment.

- **Surf 'N Store Enterprises** (located in Maui, Hawaii). Beach-based individual storage units for surfing equipment.

- **The Cocoa Lounge** (located in Anchorage, Alaska). High-end dessert lounge with hot-chocolate bar and other warm libations.

- **Active Fitness** (located in Chicago, Illinois). Fitness club for older adults, the obese, and others with health problems.

- **Student Solutions** (located in Columbus, Ohio). Low-price furniture rental for university students.

The audience. Your audience is a group of potential investors. They are smart and analytical, and they will be ruthless in assessing whether your plan for starting a new company is well conceived and well researched and can give them a good ROI. You need to present them with evidence that you have developed a product or service (your choice) that can be successfully and profitably marketed. The investors will probably not expect much financial return during your first year of operation, but they might expect a 15-to-20 percent return per year over the next few years. Consider that they might also want their original investment dollars returned to them within three to five years.

The communication strategy. Potential investors will want a business plan to meet several standards before they give it serious consideration. First, they will need to see substantial evidence for the business's likely success—not just enthusiastic claims. So your careful research and thinking will be critical. Supporting evidence does not always have to be financial or numerical; it can also include expert testimony, interviews, anecdotes, and so on. Second, investors will want to see that you have used a solid reasoning process. Clearly, you need to project possible financial and other results for your start-up company, while also demonstrating to readers why these projections are reasonable and plausible— not just wishful thinking. Third, the investors will take more seriously a business plan that is clearly written, consistent as it moves from one section to the next, and easy to follow because of its good document design.

Contents of the business plan. While there is no uniform standard for the contents of business plans, most successful plans include the sections described on the next page. Note, however, that the issues and questions explored in each section of your plan will depend to some extent on the nature of your product or service. The following descriptions are typical for many business plans, but you need to adapt them to fit *your* start-up business.

- **Cover page (1 page).** Devise an interesting logo for your proposed company and a title that clearly indicates the nature of your company and the purpose of your report. Effective document design begins with the cover page.

- **Title page (1 page).** Include the title from the cover page and also list the authors of the report and the submission date.

- **Table of contents (1 to 2 pages).** List the main headings exactly as they appear in the report and the page on which each section begins. On a separate page, list (with page references) all the tables and figures included in the report. Keep in mind that the term "tables" refers only to actual tables of numerical data. The term "figure" is used to designate all other types of visual material (graphs, charts, photos, and so on).

- **Introduction or Executive Summary (2 to 3 pages).** This section provides a succinct summary of the features of the proposed business that would be most pertinent to a prospective investor; for example, a precise description of the product or service, why and how this product or service will sell to a particular market or demographic, a quick view of competitors in this business field (and why your version of the product or service will sell successfully), the dollar amount(s) you are seeking from investors, and the projected profits to be made by investors (and over what period of time).

  If you want an actual investor to read beyond the Executive Summary or Introduction, this section must be engaging and fact-informed. While you are "selling" the business idea, enthusiasm alone will not persuade a sophisticated investor, and hype will undoubtedly kill her or his interest in your plan. Make a strong case, but don't, for example, exaggerate your anticipated market share or the short-term profits to be made.

- **Market analysis (5 to 8 pages).** Based on research, identify the likely customers or clients for your product or service. Who are they, what are their consumer interests and patterns, where are they located, and how much will they spend on your product or service? What are the trends and challenges (risks included) for this market? What legislation or regulations in effect or pending will affect your industry? Who are the major competitors, and how will you compete successfully with them? As you do research for this market analysis, consider especially the broad economic and political picture that may affect your product or service (or your potential customers), cultural trends of importance to your proposed business, environmental impact, new technologies, and energy and transportation issues that might affect your business.

- **Marketing/advertising plan (5 to 8 pages).** This section, closely allied to the market-analysis information, presents strategies for reaching and persuading your target market(s). How will you get the attention of your customers or clients in effective ways, and what is each advertising medium likely to cost? Can you advertise more effectively than your competition? How will you evaluate the success of your marketing strategies? Be sure to identify the reasons you are planning to advertise in a certain newspaper, on a particular radio station, or through some other media outlet—for example, how many people read, watch, or listen to this media source, and what are the audience demographics? Describe the "grand opening" strategies for your business as well as ongoing advertising efforts. Create at least one concrete example of an advertisement or campaign for the report.

- **Financial plan (5 to 8 pages).** The financial information will be of crucial interest to potential investors. To meet the expectations of a university writing course, you need to represent the main financial factors in clear and persuasive writing, not just in tables or figures; instead, put spreadsheets and elaborate tables or figures in an appendix. You need to research and obtain realistic figures for all major costs of starting and sustaining your business (for example, costs of office and warehouse space, of salaries and benefits, of legal assistance, of insurance, of office equipment and supplies, of production and materials, of distribution, of Web site creation and maintenance, and of utilities and other operational needs).

    The revenue parts of the financial plan are more speculative than the start-up costs you have researched but must also be realistic and based, as far as possible, on research. How many purchasers of your service or product can you realistically expect during the first year, the second, and the third? How much money do you need from the investor(s), what can the investor(s) reasonably expect for ROI, and over what period of time can they expect this ROI? Note that many investors will want a strong return—perhaps 15-to-20 percent on average per year over a three-year period—and will want also to reacquire their original investment at the end of three to five years. Can you offer a financial forecast that makes a reasonable case for such returns?

- **Management/organizational plan (5 to 6 pages).** How will the business be organized as a legal entity, and why? To answer such questions, you will need to understand, among other things, the differences between a "limited liability partnership" and an "s-corporation." How many employees do you need to begin the company (and to grow over three years), what will their responsibilities and required skills be, and how will you structure the reporting relationships? Construct an organizational chart to help your potential investors visualize the key personnel positions. What training might employees need? What will be your operating hours (and why), company philosophy, medical and retirement benefits (and costs), and retention and promotion procedures? Overall, this section should describe the key human resources you will require and how you propose to organize, treat, and manage employees.

- **Implementation plan (5 to 6 pages).** This section examines all the nitty-gritty, practical details of how the company will operate. For example, what is your time line for setting up, opening, and growing the company? (You might want to construct a Gantt chart as part of this piece—you can find resources online to help you do this.) How will you obtain needed supplies, manufacture your product or create your service, store and deliver the product or supplies, respond to customer inquiries or problems, or repair or replace your product as needed? What size and type of physical facility will your company need and at what cost? To gather this information, look on Web sites that advertise commercial real estate for lease in the community in which you would establish your business. Then try to find the rate for a facility that would meet your needs.

- **Plan conclusion (1 to 2 pages).** Here, you will summarize the most attractive and most important aspects of the business plan. This is your last "pitch" for the investor audience.

- **Works Cited (3 to 5 pages).** Following the Harvard Business School citation format or a format indicated by your instructor, this section gathers all of the works cited for the entire plan and arranges them in alphabetical order. Note that there are format differences between footnotes and the entries in a bibliography; these distinctions are clearly presented in the Harvard Business School *Citation Guide*. To access this resource, visit www.library.hbs.edu and enter "citation guide" in the search field.

- **Appendix (optional).** If you choose to provide an appendix, you might want to include the résumés of all team participants, a copy of a survey you administered to gather information for your plan, a sample advertisement for your business, financial spreadsheets, and any other information you think might be of interest to potential investors.

Further help on this project. Extracts from a student-generated business plan are included at the end of this chapter so that you can see how one team went about researching and writing an effective plan.

Bplans, at www.bplans.com, offers many short, real-world business plans on its Web site. Keep in mind that these sample plans do not illustrate the hybrid version described in this chapter. Also, use sample plans only to stimulate your own thinking and research.

Your university library might have this multivolume resource published by Gale Research: *Business Plans Handbook: A Compilation of Actual Business Plans Developed by Small Businesses throughout North America* (Detroit, 1995–).

For detailed descriptions of the *possible* content of each business-plan section, visit the Web site of the U.S. Small Business Administration at www.sba.gov, and enter "business plan" into the search field.

The following are among the most useful online databases for your research into demographic information, competing businesses in your geographic area of choice, market trends and challenges for particular types of companies, and current analyses of different businesses. Many or all of these resources may be available online through your university library:

- **American FactFinder**, at factfinder2.census.gov, is a giant warehouse for demographic information that can pinpoint the region in which you want to operate your start-up business.

- **ProQuest Statistical Insight** is a compendium of all sorts of socioeconomic and cultural statistics that can help you understand your targeted customers.

- **LexisNexis Academic and Business Source Complete** are probably the most useful of the giant databases for finding up-to-date analyses of the factors that affect the overall business climate and particular types of businesses.

- **The Social Media Examiner**, at www.socialmediaexaminer.com, provides many insights into the current use of social media to market new products and services.

A final note on research. Plain old "Googling" has earned a place in university research, but the Internet is also filled with commercial Web sites that want to sell you something along with the information they provide. In addition, many of the most thorough and objective assessments of business opportunities, trends, and challenges can be accessed only through subscription databases that a university or college library pays for and provides to its students. So get your money's worth as a student and use the databases, trade journals, academic articles, interviews, surveys, and contemporary assessments that your library offers. Many of the resources are available not only through library-based computers but also by "proxy" on your own home computer.

Student Responses to Selected Applications

Following are sample student responses to five of the previous applications: 7-B (ethics memo), 7-C (RFP), 7-D (letter of inquiry), 7-E (grant proposal), and 7-F (business plan). These documents, produced by undergraduate students at the University of California, Santa Barbara, offer insights into many aspects of the longer projects presented in this chapter. That said, the students' decisions about such things as tone, vocabulary, business and ethical values, document design, and research strategies will not always be like yours.

That's as it should be, for effective business writing is not a generic or formulaic enterprise; rather, it involves many choices made by individuals or by groups of people as they navigate different business situations within the framework of their own personalities and experiences. It's a lot easier to say what constitutes a bad business communication strategy than it is to narrow the range of possible good choices. All that being said, the student writing samples in this section will likely stimulate some new ideas for you.

Response to Application 7-B: Write an Ethics Advisory Memo

A team of students drafted the ethics advisory memo that appears later in this section (see page 191) in response to "Case 22," a fictional but realistic ethical dilemma developed by Dr. Lisa H. Newton of Fairfield University. In addition to discussing possible responses to this case, the students also viewed and discussed the video *Ethics in America II: Risk, Reward, and Responsibility* (see Application 7-B, page 167).

In Case 22, Dr. Newton presents a number of complex ethical, legal, and financial challenges. Indeed, there is no entirely clear path to an equitable resolution among the competing concerns. As you will see from the memo that follows the case, the students decided to explore several options for the CEO to consider; then, they recommended a particular course of action.

Case 22*

G. David Thorsten, company ethicist for UXL, one of the country's largest (remaining) steel producers, had just settled into his office for the morning when Tony Francato, the new hire in the legal department, came to the door in a state of obvious agitation. "Can I talk to you for a minute?"

"Of course, Tony, what's on your mind?"

"Dave, I just got a call from an OSHA man . . . uh, that's the local Occupational Safety and Health Administration inspector. He wants me to go with him on an inspection of our Rambo River coke facility. That's the plant that got the citation last year for dirty air—excessive workplace air pollution, they called it. You have to have only so much SO_2 and other junk in the air or they close the factory to protect the workers' health. And we had too much. So when I got the call I told him I couldn't talk to him for a minute, but could I get right back, and then I called Joe Salvatore, the plant manager at Rambo, and asked are we clean enough to survive an OSHA visit? He said, 'Hell, no,' excuse the French, 'we've got a very high production rate right now and the weather isn't friendly, there's a temperature inversion, the air is stagnant, and it's for sure it wouldn't pass.' Look, he thinks they'll close the plant if the inspector sees it like it is now! He said stall. Lie if you have to. There's no one else in today in my office, so I don't know if this has ever come up before and what we've done about it. Dave, can I do this?"

"I don't understand. What would you do?"

"Every OSHA inspection is accompanied by someone from the legal department and by the occupational safety manager, in this plant Bob Watson. Watson's a good man, not likely to blow any whistles. I can just call OSHA back and tell them Watson's out of town—I already told Joe I might do that, in case OSHA calls the plant to check—but he'll be back Thursday and we can go first thing in the morning."

"What good would that do?"

"By Thursday, we can have that place clean as a whistle. Cut production way down, set up fans, really blow the place out. That way we'd pass the inspection and they'd leave us alone for another year. If we fail, they're very likely to start proceedings to close the plant."

"But look, Tony, when production started up again, the conditions would be just as bad as they are now, right? And that's bad for the workers, isn't it? OSHA didn't just make up these standards out of the blue, right? OSHA or not, we have a responsibility to take care of our workers' health. Why don't we play it straight? Go through the inspection, get together with OSHA on the results, and settle on some way to clean up that air for good."

"Dave, UXL isn't going to clean up that plant. The kind of pollution-control machinery they need costs millions, and they wouldn't spend that money. You know the state of the steel industry, I guess. UXL used to be the biggest steel producer there was, but that was

*This case was prepared by Dr. Lisa H. Newton, Professor of Philosophy at Fairfield University in Fairfield, Connecticut.

before it diversified. Now it's mostly into insurance and that chain of fast-food restaurants, and . . ."

"OK, so they won't spend the money. How long is the OSHA man going to wait for you to return his call, by the way?"

"I said something about this intestinal problem I've got. No, they won't spend the money, and the way the industry is going, there aren't any more jobs out there for these workers. This plant is only marginally profitable as it is. At present levels, though, it can keep making money for another ten years at least."

"But Tony, we can't leave the men in that atmosphere for ten years, or ten weeks! Pollution kills! It causes lung cancer, emphysema, heart disease, all manner of dreadful things. I'm not sure that we wouldn't be legally liable, if one of the workers came down with lung cancer, decided it was our fault, and sued, although you'd know more about that than I would. But it's simply wrong to poison them, under the compulsion of threat to their jobs."

"Yeah, but they say they'll take that chance. The union's said they want the plant open, and so have individual workers — they won't sue. It's jobs now — food on the table, clothes for the kids, not to mention self-respect — or health later, the way it looks to them. And health later doesn't seem anywhere near as important. The way I see it, I think, it's their health and their choice. I'm not even sure that we have a right to make that choice for them. But what do you think? When I call the guy back, what do I say?"

Now let's turn to a sample advisory memo (below) that responds to Case 22 and makes recommendations to UXL's CEO about how to proceed. Note how the appointed team attempts to foreground the ethical responsibilities of UXL while also addressing the practical, financial, and legal issues that a company would need to consider. The ethical issues need to find their place — hold their own — in this complex business context.

UXL
Pittsburgh, Pennsylvania

DATE: May 16, 2016
TO: John Steele, CEO
FROM: Nick Kohan, David Love, Jennie Stodder, Scott Pantoskey, Ally Diamond
Subject: Ethics Advisory Memo

Background

On behalf of UXL, we have been assembled as the advisory team to evaluate the current situation that the Rambo River coke facility is facing. Last year, this plant received a citation for excessive air pollution and has been closely monitored by the Occupational

continued

Safety and Health Administration (OSHA) ever since. It has come to our attention that OSHA plans to do an inspection on this facility and we are concerned with the potential outcome. Joe Salvatore, the plant manager, has informed us that the plant will most likely not meet OSHA regulatory standards, therefore failing the inspection and increasing the chance of plant closure.

In order to resolve this issue and meet OSHA's regulatory standards, pollution-control machinery would have to be purchased. However, due to the current state of the steel industry, profits have been marginal and it is unlikely that the facility could afford to cover the costs of the necessary machinery installation. That being said, while business has slowed as of late, the Rambo River coke facility is projected to be profitable for at least another ten years.

Furthermore, there are concerns regarding the Rambo River employees. If employees continue to work in this polluted environment there could be negative consequences for both the workers and the company. The pollution currently in the facility could make employees subject to health risks such as cancer or emphysema. In addition, the company could face a serious legal crisis if an employee were to file suit for violating labor rights. The workers union has assured us that they would like the facility to continue operating regardless of the health issues, as many of the employees rely on the work.

After careful deliberation, our advisory team has proposed three plausible courses of action. Each of these options has been carefully outlined and is listed below.

Option 1

The first option would be for the entire company to be cleaned and restored according to OSHA standards. New machines would be purchased, and each facility would appear clean upon inspection. A companywide renovation would be costly and with UXL's marginal profits it may not be the most affordable option, but the company may not survive if the Rambo River coke facility is closed down by OSHA because Rambo River is a top-producing facility. Cleaning and restoring the entire coke plant to meet OSHA standards would be the best option for the company's reputation in the long run because it would prevent future lawsuits for employee health hazards. Employees would be able to keep their jobs if all UXL facilities were renovated, and future citations in facilities other than the Rambo River coke facility would be prevented. However, this is not the most ethical option for the company to undertake because without direct control by UXL, it is not certain that the plant will abide by the regulations. Financially speaking, this is the least practical option.

Company Impact
If new machines are purchased and facilities are cleaned, the company can advertise that it is environmentally friendly. This could boost the company's reputation as an innovative leader in the steel industry, but it could also reflect poorly on its management if word gets out about the amount of pollution that is currently being generated. In addition, it is unlikely that the company would approve spending millions of dollars to help this individual facility. Buying machines would be expensive, and the facility would have to be shut down for a few days while being cleaned. Workers would expect to be compensated for the hours they cannot work.

Employee Impact

Employees would keep their jobs after the facility is cleaned. The plant would be assured of staying open in the future, and employees would not have to worry about

getting laid off. Employees would be out of work during the few days the plant is shut down, but they may be compensated, depending on what UXL decides.

Ethical Considerations

This option seems ethical on the surface, but it depends on how UXL decides to clean up the facility and on what types of machines it chooses to buy. If the company tries to save money and cut corners by making changes that only mask the pollution taking place, this would mislead customers and employees into thinking that they are working with, or purchasing from, a company that is environmentally friendly and creates little health risk. If UXL decides to abide by the regulations and purchase less polluting machines, this would be an ethical option.

Option 2

A second possible option would be for the company to stall. By lying to the OSHA inspector in order to buy a few extra days, production could be cut down significantly, and the plant could be cleaned out and pass the impending inspection. This option would save jobs and would not require any outside funding or help from UXL as all matters could be handled within the plant. However, this option is far and away the least ethical of the three. Long-term concerns such as the employees' health would effectively be sacrificed for an immediate, short-term fix. Further, if unsuccessful in duping the OSHA inspector, the company could damage its reputation, jeopardize its relationship with its workers, and face costly litigation.

Company Impact

If everything were carried out effectively, there would be no immediate impact to the company as a whole. Production would be cut for a few days, but other than that normal operations would be sustained, and the plant would continue production with the conclusion of the inspection. However, if employees fall ill in future periods a lot of negative attention would be drawn to the company. In addition, if the cover-up were to be discovered, lawsuits and government intervention would surely come.

Employee Impact

Employees would remain employed at the plant, satisfying short-term needs. Steel workers are not very marketable, and long periods of unemployment could potentially await any of the workers employed. However, severe damage could come to these workers in the form of lung cancer, emphysema, and heart disease if they continue to work under these conditions.

Ethical Considerations

From an ethical standpoint, this would be an extremely poor choice. If somehow the cover-up could be continued for the full decade the plant is expected to yield profits, these marginal profits would be all there is to show for the workers' exposure to these unacceptable conditions. There is no transparency either, as this plan requires not only lying to OSHA but also failing to disclose the crippling effects awaiting those

continued

employees working for the plant. This option does nothing to solve the problem at hand, and it puts profits before the well-being of our employees.

Option 3

Finally, and what we are most strongly advocating, is the recommendation that the Rambo River coke facility go through with the inspection. We believe that although the facility may not be able to meet regulatory standards, the inspection is in everyone's long-term best interest. This option provides the legal and safest solution to the OSHA inspection.

Company Impact

For the company in the short term, there would be a decrease in profits because the facility, if it didn't pass the inspection, would have to be temporarily closed. Although this seems to have an immediate downside, in the long term, it is what is best for UXL. Since our company has had an honest and a reputable past, it's essential that we stick to these standards and think long term. The long-term future of UXL is what we are concerned with, and if this inspection closes the facility, we advise that the proper modifications and changes be made to maximize profits and safety.

Employee Impact

Employees would temporarily be out of work but would certainly be provided workers' compensation for their time away from the facility. Also, we advise that this option is the best for the workers because even if they were aware of the dangerous conditions they are working in, we should pursue the legal and safest option.

Ethical Considerations

From an ethical standpoint, it's clear that this option is preferable to the others. Although a temporary closure of the Rambo River facility would reduce profits temporarily, UXL is more concerned with the health and well-being of our employees. It would be unethical to simply "mask" the issue temporarily and continue to run the coke facility. This option solves the matter at hand long term and represents UXL's standards of ethical and honest decisions made on behalf of its employees.

Conclusions and Recommendations

As a team, we recommend moving forward with option 3, going ahead with the planned OSHA inspection and then working with OSHA to come up with a short- and long-term solution to the air-pollution violations. Because of our responsible past, we see no reason that OSHA would not be willing to sit down and collaborate with us on an appropriate course of action. Although we would most likely suffer temporary economic losses, it is the most ethically responsible option and it takes into consideration the financial well-being of our workers. In addition, in going with option 3, this company is choosing a long-term solution that will enable us to take advantage of the profits that we are projected to receive for the next ten years.

If you agree to move forward with this option, we promise our utmost commitment to this company and its future and will work with OSHA and management until we get this company back on its feet and restore its once-trusted reputation.

Response to Application 7-C: Write a Request for Proposals

The student who wrote the following RFP had participated in classroom discussions of the tone and content of RFPs and also viewed several online examples of RFPs. Note how carefully the student both presents the broader goals of his foundation and describes the procedures for submitting a proposal—thus balancing philanthropic ideals with financial and logistical practicalities.

21st Century Literacy Foundation: Request for Proposals

Our Mission

The 21st Century Literacy Foundation aims to increase literacy in low-income and rural areas, as well as to introduce better technology to these areas. Our foundation focuses on increasing "21st century literacy"—that is, it supports programs that aim to improve people's ability to read and write in the traditional sense, as well as their ability to communicate effectively using new technologies. Newer technologies, such as new computer/Internet software and programs for computer tablets, are revolutionizing methods of communication. In addition to offering new methods of communication that will be important in the 21st century, these new technologies can be important supplements to traditional instruction and can also offer children with different learning styles an alternative to the traditional method of learning to read and write. Therefore, our grants go to local groups focused on supporting libraries, initiating literacy and technology programs, and promoting a love of literacy.

We fund projects that

- support libraries in low-income areas, through donations to increase quantity and quality of books in circulation
- aim to introduce technology (WiFi, newer computers, tablet devices, etc.) to libraries in low-income areas
- promote rural literacy through the use of mobile libraries and related programs

Through this Request for Proposals, the 21st Century Literacy Foundation invites nonprofit organizations to apply for grants that will support initiatives to improve literacy in low-income and rural areas.

This Request for Proposals is for a one-year grant; however, successful organizations may apply to renew the grant at year's end. Grant requests can be for any amount up to a maximum of $50,000. Applications will be accepted until July 5, 2016.

---

continued

Eligibility

In order to be eligible to receive a grant from the 21st Century Literacy Foundation, organizations must meet the following criteria:

- Nonprofit organization with demonstrated 501(c)(3) tax-exempt status
- Located in or within twenty-five (25) miles of the low-income or rural community in which the proposed project is to take place
- Solid, established organizational structure with a demonstrated ability to appropriately handle grant funds
- Evidence of progress in existing projects involving "21st century" literacy development (e.g., programs demonstrated to have increased youth enrollment in library programs, increased participation in literacy programs, evidence of donated technological equipment, etc.)

Application Process

Applications will be accepted until July 5, 2016. Filing your application earlier will *not* improve your chances of receiving the grant. The grants will be awarded to the highest-quality applicants after a comprehensive review process. Applications will be accepted by mail only, and should be addressed as follows:

Michael Cipriano, Senior Grants Coordinator
21st Century Literacy Foundation
3900 State Street, Suite 220
Santa Barbara, CA 93105

You will be notified that we have received your application within two (2) weeks. Grant awards will be decided one month after the application deadline, and grant recipients will be notified of their award approximately three (3) weeks after the final decisions have been made. If you have any questions regarding the application process or want further information regarding the status of your application, please call (805) 733-3444.

Application Contents

All applications should be mailed to the address stated above and should contain the following:

- Cover sheet, including contact information for your organization
- Brief summary of your proposed project
- Mission statement of your organization
- A brief history detailing previous projects and their impact, as well as the extent of your organization's existing contributions to the community
- Detailed proposal that includes:
  1. Primary aim of your project
  2. Target community or area included in your project
  3. Projected cost of the project (amount requested for the grant)

4. Comprehensive explanations of all activities involved with the project, and a detailed budget plan, matching expected costs with planned activities
5. Names and contact information for staff members responsible for orchestrating and implementing project activities
6. No more than five (5) pages
- Conclusion (reiterating long-term project goals and target results)
- Appendixes
   1. Documentation of tax-exempt status
   2. Copies of most recent financial statements
   3. Optional, with a maximum of three: endorsements from members of the community involved in current or former projects

Response to Application 7-D: Write a Letter of Inquiry Preceding a Full Grant Proposal

A team of five students wrote the following letter of inquiry for a grant supporting Project Academic Success, a nonprofit organization that aims to provide "better opportunities for low-income students to attend four-year universities." (The full proposal for this grant is excerpted on pages 204–13.) The letter of inquiry makes evident that the students have a reasonably focused understanding of both their goals and the practical means of realizing them.

1223 Garden Street
Santa Barbara, California 93101
April 25, 2014

Jon Ramsey, Senior Grant Coordinator
New Horizons Foundation
2211 Berkshire Road
Pasadena, California 91104

Dear Mr. Ramsey:

As the Director of Development for Project Academic Success (P.A.S.), a nonprofit organization aiming to provide better opportunities for low-income students to attend four-year universities, I am interested in acquiring financial support to aid our organization's mission. We are pursuing a grant of $100,000 to cover various expenses: partial scholarships for graduating seniors who show extraordinary progress through P.A.S.; two-week training programs for the University of California, Santa Barbara, undergraduate tutors; transportation for tutors to the teaching facilities; and efforts to expand P.A.S. to various schools that currently utilize Supplemental Education Services (SES) in the Santa Barbara Unified School District but that are still underperforming, which is where we see the largest area for improvement and believe they could benefit most from the services we provide.

Introduction

Project Academic Success is a nonprofit organization that utilizes undergraduate students from the University of California, Santa Barbara, to tutor low-income high school students in the Santa Barbara Unified School District. P.A.S. focuses on revamping the SES program implemented by the No Child Left Behind Act so that socioeconomically disadvantaged students can receive the full benefit of the program and the help they need to succeed. Through our program, students are tutored by undergraduates from the University of California, Santa Barbara, who specialize in the core curriculum—mathematics, English, and writing—and are also able to cater their instructing to each student's individual needs. Our program adds SAT and ACT tutoring to supplement the core curriculum and personal mentoring in order to increase the students' chances of success. The tutors have also been trained in how to best advise students entering into the college application process, specifically on how to succeed academically even while facing emotional and economic stress at home. The program is self-sustainable as participating tutors receive compensation by means of university credit or community service hours. We believe the students at UCSB can become not only great academic advisers, but also superior personal mentors due to the prestige of the University. Through P.A.S., we believe we can fill in the holes that the government project has created.

Our program has been very successful in the two alternative schools in the Santa Barbara Unified School District that do not qualify as "program improvement" schools; hopefully, we will receive the necessary funding in order to expand our success to improving the SES in program improvement schools and create positive change in our community.

Description of Organization

Project Academic Success was founded in 2008 in order to solve problems that the government overlooked in its implementation of the No Child Left Behind Act. P.A.S. strives to provide accessible mentoring for socioeconomically disadvantaged high school students who suffer educationally from the inefficiencies in government-funded SES.

As an organization of five years, we have secured grants totaling $28,000 from the Kresge Foundation, Federal Pell Grant Program, and College Bound Foundation in order to implement our programs and change adolescents' lives. Through our work in the Santa Barbara Unified School District thus far at San Marcos High School, we have instructed 50 individuals. Not only have 46 of our pupils graduated from high school, but two-thirds of them will be attending community college. Compared to the statewide graduation rate of 80.2% for the 2012–2013 class, 92% of P.A.S. participants graduated.[1] Additionally, five students are currently attending four-year universities located in California. Last year, we were able to award two students who showed tremendous improvement with partial scholarships to continue their education with community college classes.

Our greatest success story is that of Jason Corbisez. At the age of fourteen, Corbisez' mother passed away. His family's accumulation of massive debts resulting from her medical bills as well as insurmountable grief put him in deep depression, and as he suffered, so did his grades. He signed up for Project Academic Success in his junior year of high school, in 2010. We are proud to say that today he currently is enjoying outstanding academic success at the University of California, Santa Barbara, and volunteers his time as a P.A.S. tutor.

There are several other students who, through our program, realized that with perseverance and assistance, they are capable of achieving academic success. This grant would allow us to help students like Jason, and increase the opportunities available for low-income students to succeed through higher education, despite their economic disadvantages. With additional funding, we will be able to train more tutors, expand our program to improve the high schools that are receiving SES but are still underperforming, and continue to change lives.

---

[1] California Department of Education, "Cohort Outcome Data for Class of 2012–2013," March 24, 2014, http://dq.cde.ca.gov/dataquest/cohortrates/GradRates.aspx?cds=00000000000000&TheYear=2012-13& Agg=T&Topic=Dropouts&RC=State&SubGroup=Ethnic/Racial, accessed May 1, 2014.

continued

Statement of Need

Since 2008, three of the five high schools in the Santa Barbara Unified School District have been deemed "program improvement" schools, meaning that they failed to meet Adequate Yearly Progress under the Elementary and Secondary Education Act for the two years prior.[2] Despite the fact that Supplementary Education Services (SES) have been provided through government funding in the area for five years to support academic improvement and attain Adequate Yearly Progress, high school Academic Performance Index (API) scores have continued to drop further below the California Department of Education's baseline.[3] Recent research regarding the effectiveness of SES shows flaws in the system that are pertinent to the success of the program, and therefore, the participating students.

Most of the problems that inhibit the success of SES are related to cost. Due to rising costs for tutoring service providers, the majority of the providers available to students are cost-effective online providers, which have proved to be less effective. Also, because of limited funding for SES, student eligibility to participate in the program has become more strict and therefore has excluded students who could potentially benefit from it. Accordingly, this has led to larger SES tutorial groups with fewer tutors, which allows for less instructional time per student. Lastly, SES providers lack a curriculum to follow and frequent monitoring to ensure that the curriculum is followed.[4] P.A.S. is able to resolve these inefficiencies by providing one-on-one or small-group tutoring sessions through no-cost, trained tutors. Tutors not only will be intensely trained in the curriculum by the students' teachers and have an open flow of communication to discuss each student's progress but also will be passionate in helping students to grow and reach their full potential.

In 2011, roughly 87% of high school students in California were deemed socioeconomically disadvantaged, and about 20% of those students did not graduate from high school.[5] P.A.S. improves upon the existing program by providing individualized tutoring sessions specializing in the core curriculum, as well as SAT and ACT workshops. In addition, P.A.S. hopes that the undergraduate students at the University of California, Santa Barbara, will inspire high school students to further their education, as well as serve as mentors to guide them there. Project Academic Success seeks to enhance SES programs by establishing a widely accessible, self-sustainable mentoring system that will help socioeconomically disadvantaged students in Santa Barbara county achieve academic growth and success.

---

[2]Santa Barbara Unified School District, "Program Improvement School List," May 23, 2013, http://www
.sbunified.org/districtwp/wp-content/uploads/2013/03/PI-Chart.pdf, accessed April 29, 2014.
[3]California Department of Education, "2012–2013 Accountability Progress Reporting," March 2014,
http://api.cde.ca.gov/Acnt2013/2013GrowthSch.aspx?allcds=42767864235727, accessed April 29, 2014.
[4]Burch, Patricia, Ph.D., "Supplemental Education Services under NCLB: Emerging Evidence and Policy
Issues," Education Policy Research Unit at Arizona State University, May 2007, http://www.aei.org/files/2012
/03/05/-the-implementation-and-effectiveness-of-supplemental-educational-services_17150915643.pdf,
accessed April 29, 2014.
[5]California School Boards Association, "California High School Dropout and Graduation Rates 2010–2011,"
May 2013, http://www.csba.org/GovernanceAndPolicyResources/~/media/CSBA/Files/Governance
Resources/GovernanceBriefs/201305FactSheetHSGradRates.ashx, accessed April 29, 2014.

Methodology

Our tutors, who apply and are then chosen by the P.A.S. staff, are trained at the P.A.S. headquarters by the students' teachers from Santa Barbara High School, the school we plan to expand to, through an intensive two-week training. Tutors are trained to teach curriculum by the high school teachers and will report to the teachers throughout the program on their student's progress. The tutoring sessions are usually an hour in length and take place twice a week on the student's campus. As the tutors complete the training and tutoring, P.A.S. grants UCSB students either the equivalent amount of community service hours to the time they spend teaching or possibly two units per quarter, pending approval by the University of California, Santa Barbara.

Our educational services include private tutoring in reading, writing, and mathematics; standardized test preparation; and workshops. For the standardized test preparation, P.A.S. offers personalized tutoring for standardized tests such as SAT, ACT, AP, and SAT Subject Tests. Students learn about the strategies for approaching different question types as well as test-taking strategies for the SAT and ACT.

P.A.S. headquarters is located in Santa Barbara. At San Marcos High School, our current program location, tutoring classrooms have been designated for our programs. We provide one-on-one or small-group tutoring services to students specifically from low-income areas. For those qualified students, we carefully match them with their own tutors who report their progress to the teachers. The tutors also provide suitable academic guidance regarding college applications and methods for taking the college admission exams. Our tutors are successful in aiding with this process having just gone through the process recently themselves. Our core curriculum mainly consists of reading, writing, and mathematics, but students can request additional help on specific subjects. In addition, we hold weekly workshops on campus every Saturday as a group study for the SAT and ACT, encouraging students to participate in open discussions about different problems they have encountered. Some of the graduating seniors who show vast improvement through our program and potential to grow further receive partial scholarships by P.A.S. to help them continue their studies in community colleges, universities, and vocational schools.

Other Funding Sources

We have also secured partial funding for our program from three foundations. We have obtained $17,000 from The Kresge Foundation, $10,000 from The Federal Pell Grant Program through the U.S. Department of Education, and $7,000 from The College Bound Foundation. All three of these foundations promote equal opportunities for students, regardless of socioeconomic status, and promote the value of education.

continued

Summary

Project Academic Success (P.A.S.) believes that every student deserves the same opportunity to succeed in academia, regardless of his or her socioeconomic status. Our organization offers tutoring, as a means of supplementing the core curriculum, to students who qualify as low-income in the Santa Barbara Unified School District, specifically in schools that are deemed "program improvement" schools by the No Child Left Behind Act and that offer government-funded tutoring services that are not fulfilling the students' needs. This is accomplished by bringing trained tutors from the University of California, Santa Barbara, into the schools, rather than low-income students having to find transportation to an outside facility. In addition to tutoring, P.A.S. believes that it is important to nourish the spirit as well as the mind; therefore, we also provide mentoring to these young men and women. This mentoring supports students while in high school and prepares them for life after graduation by making them aware of their options and helping them in planning their future so that they may achieve their goals. We appreciate New Horizons' time and consideration for supporting our program and helping to provide our students with the necessary tools to reach their fullest potential. If you have any questions, please contact Martha Grimes, Director of Development, at (760) 504-4184.

Sincerely,

Martha Grimes (representing Laura Francis, Jenna Thompson, Daniel Levens-Lowery, Xiangdi Wang)
Director of Development
marthagrimes@projectacademicsuccess.com
(760) 504-4184

Response to Application 7-E: Write a Grant Proposal

The students who wrote the letter of inquiry for the Project Academic Success grant (see the previous section) also wrote a full grant proposal for the project. Because the proposal is quite extensive, presented here are just a few excerpts, including the Executive Summary, a portion of the Statement of Need, and a portion of the Budget section.

Notice that the actual pages from the proposal illustrate how the students effectively incorporated elements of document design. These elements include clear headers to help readers navigate the proposal, as well as plenty of white space between sections to provide clear transitions from one topic to the next and to give the text (and readers) breathing space.

In the Executive Summary, following the title page that's shown on the next page, the students clearly preview the issues that later sections of the proposal will address in greater depth. They also keep the summary focused and engaging, increasing the chances that readers will keep turning the pages to learn more.

Project
Academic
Success

Inspiring Students to Aspire

Martha Grimes
Laura Francis
Jenna Thompson
Daniel Levens-Lowery
Xiangdi Wang

Project Academic Success

1.0 Executive Summary

In 2001, the No Child Left Behind Act was put into effect to modernize the Elementary and Secondary School Act of 1965 that aimed to create equal opportunity for quality education in United States public schools. This legislation approved the allocation of Title 1 funds to schools where 40% of students come from government-defined "low-income families."[1] These funds were meant to hold schools accountable for providing quality education as well as provide additional educational support to students. This additional educational support comes in the form of "Supplemental Educational Services," or SES, for schools deemed in need of "program improvement" based on below-acceptable test scores. Although the idea of government-funded tutoring is ideal in concept, its years as an active program have shown immense complications and untapped potential through unchanging, below-base-level test scores.

Project Academic Success (P.A.S.) is a nonprofit organization that seeks to fill the gaps in current SES by utilizing undergraduate students from the University of California, Santa Barbara, to tutor low-income high school students in the Santa Barbara Unified School District. P.A.S. focuses on revamping the SES program implemented by the No Child Left Behind Act so that socioeconomically disadvantaged students can receive the full benefit of the program and the help they need to succeed.

Through our program, undergraduates from the University of California, Santa Barbara, travel to high schools and tutor students in the core curriculum, mathematics, English, and writing, as well as in subject areas based on individual student needs. Our program also features SAT and ACT tutoring to supplement the core curriculum and personal mentoring in order to increase the students' chances of success. Tutors become well equipped with tutoring skills and review the curriculum through a two-week intensive training course led by high school teachers of each discipline. The tutors have also been trained in how best to advise students entering into the college application process, specifically on how to succeed academically even while facing emotional and economic stress at home. The

---

[1] U.S. Department of Education, "Improving Basic Programs Operated by Local Education Agencies," June 2014, http://www2.ed.gov/programs/titleiparta/index.html, accessed May 10, 2014.

continued

Project Academic Success

program is self-sustainable, as participating tutors receive compensation by means of university credit or community service hours. We believe the students at UCSB can become not only great academic advisers but also superior personal mentors because of the prestige of the University. P.A.S. is a dual-purpose program. Although intended to provide educational support for high school students, the undergraduate experience of the UCSB students is enhanced by receiving valuable hands-on teaching and mentoring skills.

Through P.A.S., we strive to enhance SES and create productive members of society by providing accessible support and equal opportunity to socioeconomically disadvantaged high school students. With the skill sets we provide for the students, we hope to not only see an increase in college enrollment but also see the students thrive in higher education. Along with improvement in each student's individual GPA, we aim to raise the average GPA of the school. Finally, as products of higher education, we hope the University of California, Santa Barbara, undergraduates will inspire the students to aspire.

In the Santa Barbara Unified School District, all three of the traditional high schools have been deemed "program improvement" schools. As an organization of five years, we have already made a significant impact on one high school in the district, San Marcos High School, through our initial funding of $28,000 from various education-based foundations. Of the 50 students who participated in our program, 46 students graduated from San Marcos High School, which is higher than the statewide graduation rate. Two-thirds of these students are attending community college, and eight students are now attending four-year universities in California. With our funding, we were also able to provide partial scholarships to a few students who showed tremendous improvement and promise. We are so proud of our achievements thus far and are excited to better the community through further expansion.

In order to continue this progress, Project Academic Success would like to expand to Santa Barbara High School beginning in the 2014–2015 school year. Santa Barbara High School has been in need of "program improvement" for more than five years and has held this status longer than any other high school in the district. With a similar size and demographic to San Marcos High School, we feel we can implement our program and continue our successes quite smoothly.

3

Project Academic Success

With this expansion comes the need for more initial funding; however, it is mainly a self-sustainable program. We ask that the New Horizons Foundation, as a substantial supporter of innovative, proven programs, please consider us for a grant of $100,000 to cover various expenses that will inevitably improve our program and, accordingly, the community. These expenses will allow for tutor transportation to tutoring sites, field trips to the University of California, Santa Barbara, stipends for training-program teachers, rent, salaries to a few full-time employees, supplies for students, as well as increased scholarships. By having these expenses covered by a grant, we are able to keep our program free and accessible to students as well as create more benefits. Thank you for your consideration of our request.

continued

In this next excerpt, the first two pages of the Statement of Need, the students present a more detailed argument for funding Project Academic Success. Notice how they effectively weave in data from outside sources.

Project Academic Success

2.0 Statement of Need

The No Child Left Behind (NCLB) reform was perhaps the most important legislation regarding public education in 35 years. It was intended to provide improvement on a national level to our schools and for future generations. Previously, the responsibility of public education reform was in the hands of the individual states; this changed with the implementation of NCLB. The educational policy at the federal level has mandated performance standards, and the ability to reduce federal funds to states that do not make progress and to reward states that do meet these standards.[2] These reforms require that schools: adopt rigorous content standards; use scientifically based instructional practices; guarantee "highly qualified" teachers for all students and "demonstrate academic improvement by meeting pre-established benchmark percentages of proficiency in reading and math as measured through state assessments."[3] The benchmark for all schools is to have 100 percent proficiency in reading and mathematics by 2014. This is to be attained by schools meeting a pre-established rate of growth.[4] The schools that are not performing up to these mandated performance standards are labeled as Program Improvement (PI) schools. These schools would qualify for Supplemental Educational Services (SES), which would provide tutoring to low-income students to help improve their scores. Unfortunately, this program has limited success and schools rarely, if ever, make it off the PI schools list.

Since 2008, three of the five high schools in the Santa Barbara Unified School District have been deemed "program improvement" schools, meaning that they failed to meet Adequate Yearly Progress under the Elementary and Secondary Education Act for the two years prior.[5] Despite the fact that Supplemental Education Services (SES) have been provided through government funding in the area for five

---

[2] Maleyko, Glenn, "The Impact of No Child Left Behind (NCLB) on school achievement and accountability," Ph.D. diss., Wayne State University, 2011, *Academic Search Complete*, Web, accessed May 4, 2014, p. 1.
[3] Sandoval, Patricia G, "Allocation of educational resources to improve student learning: Case studies of California schools," Ph.D. diss., University of Southern California, 2009, *Academic Search Complete*, Web, accessed May 4, 2014, p. 1.
[4] Ibid., p. 1.
[5] Santa Barbara Unified School District, "Program Improvement School List," May 23, 2013, http://www.sbunified.org/districtwp/wp-content/uploads/2013/03/PI-Chart.pdf, accessed April 29, 2014.

5

Project Academic Success

years to support academic improvement and attain Adequate Yearly Progress, high school Academic Performance Index (API) scores have continued to drop further below the California Department of Education's baseline.[6] Recent research regarding the effectiveness of SES shows flaws in the system that is pertinent to the success of the program and, therefore, the participating students.

Project Academic Success (P.A.S.) was founded in 2008 in order to solve problems that the government overlooked in its implementation of the No Child Left Behind Act. P.A.S. strives to provide accessible mentoring for socioeconomically disadvantaged high school students who suffer educationally from the inefficiencies in government-funded SES. Through our work at San Marcos High School, in the Santa Barbara Unified School District, we have instructed 50 individuals. Not only have 46 of our pupils graduated from high school, but also two-thirds of them will be attending community college. Compared to the statewide graduation rate of 80.2% for the 2012–2013 class, 92% of P.A.S. participants graduated.[7] Additionally, eight students are currently attending four-year universities located in California. Last year, we were able to award two scholarships to continue their education with community college classes.

However, most of the problems that inhibit the success of SES are related to cost. Because of rising costs for tutoring service providers, the majority of the providers available to students are cost-effective online providers, which have proved to be less effective. Alternatively, the tutoring services that are acceptable for the SES program are off campus, resulting in students needing to find transportation to and from tutoring sessions, which also requires parental involvement. Furthermore, SES providers, whether they are online or off campus, lack a curriculum to follow and lack frequent communication with the teachers to ensure student progress.[8] Finally, because of limited funding for . . .

---

[6] California Department of Education, "2012–2013 Accountability Progress Reporting," March 2014, http://api.cde.ca.gov/Acnt2013/2013GrowthSch.aspx?allcds=42767864235727, accessed April 29, 2014.
[7] California Department of Education, "Cohort Outcome Data for Class of 2012–2013," March 24, 2014, http://dq.cde.ca.gov/dataquest/cohortrates/GradRates.aspx?cds=00000000000000&TheYear=2012-13&Agg=T&Topic=Dropouts&RC=State&SubGroup=Ethnic/Racial, accessed May 1, 2014.
[8] Burch, Patricia, Ph.D., "Supplemental Education Services under NCLB: Emerging Evidence and Policy Issues," Eduation Policy Research Unit at Arizona State University, May 2007, . . .

In this next excerpt, from the Budget section of the grant proposal, the students don't just provide a bunch of numbers to illustrate Project Academic Success's costs and needs. Instead, they tell the story behind the numbers, thoughtfully justifying every request they are making.

Project Academic Success

5.0 Budget

Since its inception five years ago, Project Academic Success has been able to secure $28,000 from multiple different sources in order to fund various expenses that we have come across through the function of our program. These include renting the office space for our headquarters, providing partial scholarships for outstanding students, training and transporting tutors, and providing supplies to aid in tutoring. Thus far, we have been able to get by with the funding that we have received, and we have seen tremendous results in the rising graduation rates of our students. Our success at San Marcos High School has shown how valuable our program can be at other schools in the area. As we look to expand to Santa Barbara High School, we feel that we will need additional funding in order to continue providing the services that we have been providing to San Marcos High. With that being said, we are asking for a grant of $100,000 from New Horizons so that our services can not only continue at San Marcos High, but also expand to Santa Barbara High, and hopefully to additional schools in the area.

To date, we have hired 30 tutors who have each instructed either one or two students. Our ultimate goal is to be able to support enough tutors so that each will be able to focus on one single student; increasing individual tutoring will strengthen the mentor aspect of the student-tutor relationship rather than purely the academic aspect. Additional funding will go toward increasing scholarships, additional transportation and training for added tutors, publicity, supplies, rent, salaries, and other unforeseen expenses that come up.

5.1 Training Costs

UC Santa Barbara students are recruited to become tutors for P.A.S. through a brief interview process. If we deem the student is up to our program's standards regarding academic achievement and personal mentoring ability, then they are admitted into the two-week training program at our headquarters. Since the training is concurrent with the school year, the training sessions are held Monday through Friday, 6 p.m. to 9 p.m. to avoid class conflicts. The teachers from Santa Barbara High will receive a small stipend of $500 for aiding in the two-week training process. We plan to have two trainings per year so that the tutor turnover remains reasonably long.

23

Project Academic Success

Since the tutors already know the academic aspect of the tutoring, limited supplies are needed during training. It is more focused on mentoring, specifically on which aspects of academic assistance the tutor should focus. Therefore, the training cost amounts to purely paying the teachers; the projected training cost for 2015 is $5,000 (5 teachers per training, with two planned trainings).

5.2 Transportation

Transportation during the training program is up to the tutor to arrange; however, the tutors are reimbursed through an expense report at the end of the program. For the future tutor sessions that will take place at Santa Barbara High School, we will provide vans that pick up the tutors from UC Santa Barbara and transport them to and from the high school. The round-trip mileage is 12 miles, meaning the cost per week to transport tutors will be around $50 (assuming the reimbursement rate of $0.57 per mile,[20] two tutor sessions per week per tutor, and two rotations of tutors per week). Transportation expense for 2015 is projected to be $2,000, assuming there will be time taken off for summer and other various holiday breaks.

5.3 Publicity/Advertising

For the first five years, P.A.S. has not necessarily needed extreme publicity, considering the program has been limited to San Marcos High. At the start of our nonprofit, we were able to secure funding from various foundations without doing much advertising. Our services were directly provided to the high school and our $28,000 in grants has been able to support our expenses. However, in order to expand to Santa Barbara High and beyond we would like to secure advertising space in local newspapers/magazines such as *The Montecito Journal* and *The Santa Barbara Independent*. This will aid in our quest to secure future funding for the upcoming years to continue outside sources to finance our expenses. We plan to budget $2,000 for half-page ads in a couple of local newspapers.[21] We also will

---

[20] Internal Revenue Service, "2014 Standard Mileage Rates," December 9, 2013, http://www.irs .gov/2014-Standard-Mileage-Rates-for-Business,-Medical-and-Moving-Announced, accessed May 2014.
[21] Gaebler Ventures, "Costs for Advertising in Newspapers," February 21, 2001, http://www .gaebler.com/Newspaper-Advertising-Costs.htm, accessed May 2014.

continued

Project Academic Success

set aside $450 to buy one-inch column advertisements in the UCSB *Daily Nexus* every week of the 33-week school year.[22] These advertisements aim to attract tutors from the student body, rather than targeting future funders, as the local newspaper advertising will do. Publicity and advertising make up $2,500 of the 2015 budget.

5.4 Rent

The P.A.S. headquarters is located in downtown Santa Barbara, at 928 Carpinteria St. Our office space is approximately 800 square feet, causing our rent to be $1,200 each month; and $14,400 for the year.[23] The office-building rent includes utilities and Internet, so we do not have to expense them separately. Our tutoring takes place at the school of the students, and both San Marcos High and Santa Barbara High have allowed P.A.S. to use vacant classrooms after school at no expense. This is extremely valuable to our organization because it allows us to use our funds in places that are more necessary.

5.5 Salaries

We currently have a Director of Development, Martha Grimes, who is a full-time employee at our headquarters. She coordinates the tutors with prospective students, arranges transportation, and communicates with the teachers at the students' high school. Currently, the other members of staff are part-time volunteers, and the tutors receive either community service hours or unit credits (pending approval of the university). With additional funding, we hope to expand our staff to include a university recruiter who will be able to . . .

---

[22] The Daily Nexus, "Rates and Services," January 1, 2014, http://dailynexus.com/advertising /rates-and-services/, accessed May 2014.
[23] Loopnet, "928 Carpinteria St," August 1, 2013, http://www.loopnet.com/xnet/mainsite/listing /Profile/Profile.aspx?LID=18291777&PreviousLinkCode=10850&PreviousSourceCode=1lww 2t006a00001&&LinkCode=10850&SourceCode=1lww2t006a00001, accessed May 2014.

In this final excerpt from the Budget section, notice how the pie chart makes the proportions of the projected budget instantly clear.

Project Academic Success

The breakdown of the projected 2015 Budget is shown in Figure 5-2:

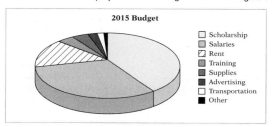

2015 Budget

☐ Scholarship
▦ Salaries
▨ Rent
▦ Training
▦ Supplies
▩ Advertising
☐ Transportation
■ Other

Figure 5-2: 2015 Budget

5.9 Future Funding

Because of the nature of our expenses, we must be able to secure future funding in order to be self-sustainable. As our organization gets additional publicity and is able to expand further across the southern part of California, we believe funding will be easier to obtain. However, in the upcoming years, we have plans to raise awareness in order to receive grants from more sponsors. We have already planned a banquet this summer in the front ballroom of the Doubletree Fessparker, where we will inform residents of Santa Barbara and Montecito about the goals of our organization. We hope to raise money through voluntary donations and the $100 event-entry fee. The Doubletree recently has decided to sponsor our organization, and is providing the ballroom at no cost for us to hold our event. They have also agreed to allow us to hold future fund-raising events at no cost as a part of their sponsorship.[29] Our biggest success story, Jason Corbisez, has agreed to speak at the banquet and share his story of how P.A.S. helped him to accomplish his wildest dreams. After being able to receive the services of P.A.S. and receiving our scholarship for the year, he was able to attend UCSB on financial aid and is currently studying to be a high school teacher. We hope that by showing a real-life success story and proving . . .

---

[29] Montecito Bank & Trust, "Monticeto Bank & Trust Business Event Series Surpasses Expectations," January 2014, https://www.montecito.com/Articles/Article.aspx?id= montecito-bank-trust-business-event-series-surpasses-expectations, accessed May 2014.

Response to Application 7-F: Write a Business Plan for a Start-Up

A team of five students wrote a business plan for AgroFresh, which aims to "promote a culture of sustainable living in the community." Specifically, AgroFresh plans to partner with Santa Barbara farms to supply local restaurants with fresh organic produce; to sell, through its store, organic produce to the larger community; and to provide lessons on urban farming to area residents.

Business Plan

AgroFresh

Anne Holston, Axle Wartanian, Kara Gorman, Nico Tomei, Maher Zaidi

Like the grant proposal excerpted in the previous section, this business plan is also quite extensive. Specifically, it comprises an Executive Summary, an Industry Analysis, a Market Analysis, a Marketing Plan, a Management and Operational Plan, a Financial Plan, References, and an Appendix. Following are selected excerpts from the AgroFresh business plan.

Let's turn first to the Executive Summary. As you read it, notice how the students incorporated data about the growing interest in organic and locally farmed produce. It will be clear to any potential investor that the students did their research.

Executive Summary

AgroFresh exists to create and deliver the freshest produce available locally and to promote a culture of sustainable living in the community. By partnering with local organic farmers and teaching sustainability, *AgroFresh* enriches the local community.

AgroFresh's business is in high demand for reasons ranging from supporting local farms to concerns about exposure to toxins in non-organic produce. Recent studies show that approximately 43% of fruits and vegetables purchased are organic, and this trend is on the rise. This translates to restaurants as well, which, on average, purchase 80% of the produce they use from local farms, and 60% of that is organic.

To provide consumers with locally grown food, *AgroFresh* partners with local farms to supply Santa Barbara restaurants with organic, fresh-from-the-farm produce. In addition, *AgroFresh* has a storefront where produce grown in the *AgroFresh* garden will be available to the local community for purchase. To promote sustainable living within the community, *AgroFresh* provides lessons on urban farming for the everyday consumer.

AgroFresh will market itself as a center of sustainability in the urban farming culture of Santa Barbara. We will market *AgroFresh* online utilizing our personal blog Web site as well as social media such as Twitter, LinkedIn, and Facebook. *AgroFresh* will be marketed in print, utilizing sustainable, recycled printing, as well as in articles in local Santa Barbara newspapers. We will also market *AgroFresh* in the community by sponsoring community events.

AgroFresh is based on the operational structure of five workers: sales manager, marketing consultant, grower/farmer, grower's assistant, and the sales associate. Workers are designed to have their own tasks at hand as well as helping the others with their positions. Everyone working together will make the company successful because involvement from all positions is required for the growth of *AgroFresh*. *AgroFresh* will endure in sustainable acts and provide customers with the highest quality in produce.

AgroFresh maintains a $500,000 start-up cost with most of that capital funding the building and employee salaries. Monthly revenue of $14,200 from selling produce and fees for urban farming classes set *AgroFresh* on track to be profitable before the end of our third year operating. With an investment in *AgroFresh*, investors can plan to make a return on their investment in just three years and from then on our business only plans to grow even greener.

Next, let's look at a page from the AgroFresh business plan that makes especially effective use of document design. In the following excerpt from the Market Analysis section, notice how the authors incorporate graphical data to show an important environmental benefit of their business. (The students used in-text citations instead of footnotes in their business plan. Full publication information appears later, in a References section.)

Reducing Fuel Consumption

In 2002, food traveled an average of 2,000 miles from farm to table (Lazaroff, 2002, para. 1). As of 2013, food traveled 1,500 miles on average (Barrett, 2013, para. 1). While the average distance food has traveled has dropped in the past 11 years, *AgroFresh* helps to cut down on the amount of fuel used even more. Figure 2 shows the average distance different produce travels in the United States. Every type of food on the graph is grown by at least one of the farms *AgroFresh* has a contract with, meaning each item of produce is grown within 50 miles of *AgroFresh*. Because the produce is traveling no more than 50 miles, *AgroFresh* reduces the amount of fuel used and does its part to keep the environment clean.

Figure 2: Average Distance from Farm to Plate (miles)

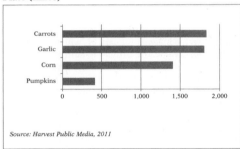

Source: *Harvest Public Media, 2011*

Additionally, *AgroFresh* promotes sustainability by providing participants who take classes with kits to start their own garden. *AgroFresh* provides seeds for customers to plant different vegetables. Once they harvest their first crops, if they save their seeds they will be able to plant more the following season. This process allows people to continuously grow food, and reduces the amount of seeds they must buy at a store. By not purchasing seeds, customers reduce the amount of fuel used to travel to the store, in addition to the fuel used to transport seeds long distances, just as produce is often transported. *AgroFresh* encourages customers to purchase our food because it is locally grown, though, should they prefer other distributers, *AgroFresh* always encourages customers to buy locally.

Finally, let's look at a page from the Management and Operational Plan. Notice how the authors use a graphic to make AgroFresh's organizational structure clear.

Overview

AgroFresh thrives in the market of urban farming. The company has a team that provides an innovative take on creating sustainable living as well as healthy eating. The employees of the company are experienced in their respective fields and contribute to the overall well-being of the company. *AgroFresh* is designed to create a team-building atmosphere by incorporating the different areas together. By working together, the employees of *AgroFresh* ensure the goal of promoting a more sustainable environment.

Internal Structure

Figure 1 Displays *AgroFresh* Organizational Structure — The structure of the company is designed in a way to ensure that leadership is present but also works closely as a team. All members of the staff are expected to work together to promote the goal of the company.

Figure 1: Organizational Structure

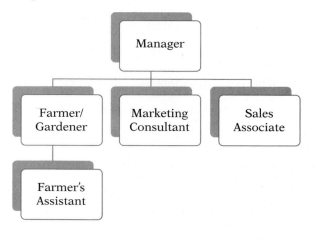

Business Writing Gaffes in the Real World

This book is filled with examples of business writing that are clear in their purpose, tone, and audience. You can learn a lot from these best practices. However, even highly experienced business professionals make communication errors at times, and you can learn from these mistakes as well.

This chapter of *Business Writing Scenarios* gathers ill-considered business communications that had consequences for the authors, sometimes very serious consequences. Most of the gaffes involve digital communications, which, as discussed in Chapter 6, can reach unintended audiences far more easily than the mistakes made on paper in years past. That's an important change for you to keep in mind: any momentary errors of judgment that you make in e-mails, tweets, text messages, or Facebook posts can be shared quickly with thousands of unintended recipients, and these communications can be subpoenaed easily for court proceedings as well.

Reviewing Key Causes of Writing Gaffes

Many of the gaffes described in the next section grew out of these typical communication problems:

- **Failing to understand the audience.** The writer does not consider sufficiently the sensitivities or suspicions of his or her audience. This can lead to a host of problems in tone and word choice, and to cloudy strategic thinking.

- **Writing and sending in a state of stress.** The writer composes and sends the communication while angry or excessively nervous, rather than stepping back to reconsider the situation and message at a later time.

- **Failing to anticipate an unintended audience.** The writer may reach an unintended audience by mistakenly clicking "reply all" or by sending a message to the wrong recipient list. Or he or she may not consider that the communication might be leaked to the press or subpoenaed for a legal action.

- **Trying to misrepresent or cover up the truth.** The writer attempts to mislead an audience, sometimes by blaming others, by denying any knowledge of the issue, or by trying to "spin" the issue when the original deception or error is exposed.

- **Making poor writing choices that can be misinterpreted.** The writer doesn't think carefully about the clarity of the communication and uses words that convey the wrong tone or content, perhaps even using words that sound racist, sexist, or otherwise insensitive.

Most of the errors gathered in this chapter could have been prevented if the writer had embraced the leadership qualities of fairness, respect, and honesty described in Chapter 9. For example, if a chief of police had considered the sensitivities of his overweight staff, he might have rethought his "jelly-belly" e-mail (see page 222) and not gotten fired. Also, if the CEO of a peanut company had been more honest and ethical about problems with his company's products, he might have avoided congressional scrutiny and disgrace (see page 226).

Touring a Gallery of Gaffes

Few of us escape making errors in our professional communications. However, we can try to be more careful in our own writing and also learn from the mistakes of others. The following sections explore some real-world gaffes to avoid.

Getting the Job the Wrong Way

Let's start with a challenge that you may confront at the very beginning of your career: the desire to present yourself in the best possible light to a prospective employer or graduate school. Are you tempted to tweak a GPA higher, to claim a degree or certificate that you did not quite complete, or to puff up a routine work experience to make it sound more impressive? If you are, keep in mind that such behavior not only is ethically wrong but also can lead to serious problems during an interview—or even to career-ending consequences.

For example, if you claim "fluency" in Spanish on your résumé when you have only a limited conversational ability, what happens when a native Spanish speaker starts speaking Spanish to you during the interview?

If you present one of your central work experiences as "traffic-flow manager," and during the interview it becomes clear that you parked cars for a hotel, your exaggeration will not impress a prospective employer.

If you falsely claim expertise with data-management software because the job requires that skill, you are likely to be found out, and perhaps fired, if you get the job and can't do the required work.

The simple truth is that your cohorts and employers will not quite trust you if they discover a recent or even a long-ago falsehood on your résumé (or posted on your blog or Web site). The initial boost you might achieve in a competitive applicant pool poses a big risk to your credibility later on.

It's hard to determine just how many people falsify résumé and cover-letter information, but according to a 2004 study reported by *USA Today*, statistics hover around 50 percent of job applicants. Many statistics about such dishonesty are gathered by commercial résumé-screening services, so they may be inflated. At the same time, it is clear that many more companies are now conducting some degree of background checking. In 2012, the *Atlanta Business Chronicle* reported that background-checking services had become a $1 billion annual business sector in the United States. With so many employers on guard, it makes practical (as well as ethical) sense to represent your skills, academic credentials, and job-related experiences honestly and accurately.

Here are a few notable cases of falsified credentials:

Kicked out of Harvard

Adam Wheeler gained admission to Harvard University in the fall of 2007 through multiple falsehoods on his application materials and eventually, in 2010, pleaded guilty to larceny and identity fraud. Even though he received 10 years of probation and a suspended sentence of two years in jail, he allegedly continued to falsify his background as he sought employment in a tough job market, as reported by the *Boston Globe* in 2010.

Dismissed as Admissions Dean

In 2007, Marilee Jones, the widely respected dean of admissions for the Massachusetts Institute of Technology, resigned from her position when it was discovered that she had faked degrees from various colleges when she applied 28 years earlier for her first job at MIT. As reported by the *New York Times* in April 2007, Jones simply allowed her lie to continue as her career progressed: "'I misrepresented my academic degrees when I first applied to M.I.T. 28 years ago and did not have the courage to correct my résumé when I applied for my current job or at any time since,' Ms. Jones said in a statement posted on the institute's Web site. 'I am deeply sorry for this and for disappointing so many in the M.I.T. community and beyond who supported me, believed in me, and who have given me extraordinary opportunities.'"

Fired as CEO of Yahoo! Inc.

In May 2012, after serving only four months as CEO of Yahoo!, Scott Thompson was forced to resign when it was discovered that he had never earned a bachelor's degree in computer science from Stonehill College, as he had claimed. As reported in the *Los Angeles Times*, Thompson did graduate from Stonehill, but the college didn't begin granting computer-science degrees until several years after he had left.

Insulting Co-workers

In 2006, the Winter Haven, Florida, chief of police, Paul Goward, sent the following memo to his officers; it quickly led to his firing by the City Council. Goward's probable intent was to encourage better health and fitness among his troops, but he crafted a memo that instead insulted many police officers in its tone and word choices. He might have been more effective by not invoking the "fat cop in a doughnut shop" stereotype and by developing a fitness program to aid his officers. (Goward claimed later that such a program was actually in the works.)

The insensitive memo was reported by Merissa Green in the *Ledger*, Winter Haven, Florida, on October 24, 2006.

Subject: Are you a Jelly Belly?

As I look around the department I see a disconcerting number of us that appear physically challenged with obesity and/or a general lack of physical fitness. This is a tremendous concern to me because the literature, to say nothing of common sense, states that if you are obese and/or out of shape you are a predictable liability to yourself, your family, your partner, this department, the city of Winter Haven, and the citizens of our city. So, take a good look at yourself.

If you are unfit, do yourself and everyone else a favor. See a professional about a proper diet and a fitness training program, quit smoking, limit alcohol intake, and start thinking self-pride, confidence and respectability. And stop making excuses for delaying what you know you should have been doing years ago. We didn't hire you unfit and we don't want you working unfit. Don't mean to offend, this is just straight talk. I owe it to you.

Here's a further tip: whenever you are tempted to write "don't mean to offend" or a similar preemptive apology, you can be sure that you have in fact written something offensive. So step back and reconsider your communication.

Issuing an Apology without Acknowledging Any Responsibility

In 2007, the New Era Cap Company, which holds clothing franchises with Major League Baseball and with the National Football League and the National Hockey League, was faulted for selling souvenir Yankees baseball caps that, to all appearances, included recognized gang signs and colors. Many people protested that this practice endangered the children to whom the caps were being marketed, and New Era quickly stopped selling them. As reported in the *New York Times* on August 25, 2007, the very brief New Era statement below is a masterpiece of obfuscation:

New Era said it was surprised by what the cap designs signified:

"Recently, it has been brought to our attention that some combinations of icons and colors on a select number of caps could be too closely perceived to be in association with gangs," the company said in a statement. "In response, we, along with Major League Baseball, have pulled those caps."

The New Era statement suggests that the issue with the caps was entirely a matter of happenstance visual associations and misperceptions by the public, rather than a bad marketing decision at New Era.

The Yankees also quickly issued a statement. This statement, on their Web site, was more straightforward than New Era's:

> The New York Yankees were completely unaware that caps with gang-related logos and colors had been manufactured with the New York Yankees logo on them. These caps were made under a licensing agreement between New Era and Major League Baseball and were not subject to the Yankees' approval nor shown to the New York Yankees at any point prior to their retail distribution.
>
> The New York Yankees oppose any garment that may be associated with gangs or gang-related activity. Upon learning of the existence of these caps this morning, the New York Yankees contacted Major League Baseball. We were notified by the Commissioner's Office that steps had already been taken to recall the caps from all points of sale. The league ensured that no such product will be manufactured in the future.

Correcting the Record . . . Sort of

In 2010, the pharmaceutical company Genentech made remarkable claims for its osteoporosis drug Boniva: "After one year on BONIVA, 9 out of 10 women stopped and reversed their bone loss." The ads, which appeared in various print publications, were endorsed by the well-known actress Sally Field.

RALLYWITHSALLY
FORBONEHEALTH[SM]

"I take BONIVA to help me manage my osteoporosis and strengthen my bones."

Like Sally Field, you, too, can help protect your bones with once-monthly BONIVA.

▸ Get started with a free trial offer for BONIVA.

In January 2011, the Food and Drug Administration (FDA) admonished Genentech, saying that the company's advertising "misleadingly overstates the efficacy of Boniva." The FDA required Genentech to withdraw all of its misleading advertising. And in the fall of 2011, Genentech published this curious retraction in various media outlets:

> You may have seen an ad about BONIVA for the treatment and prevention of postmenopausal osteoporosis that may have given you the wrong impression. Our ads stated that "After one year on BONIVA, 9 out of 10 women stopped and reversed their bone loss." The FDA has found that there is not enough evidence to support this statement and wants us to clear up any misunderstanding you may have had about these ads and make sure you have the correct information about BONIVA. BONIVA has not been proven to stop and reverse bone loss in 9 out of 10 women and is **not** a cure for postmenopausal osteoporosis. BONIVA has been shown to help increase bone mass and help reduce the chance of having a spinal fracture (break). We encourage all patients to discuss their treatment with their healthcare provider. Only your doctor can determine if BONIVA is right for you.

Notice that the retraction is forthright up to a point but also suggests that the error is a "misunderstanding" on the readers' part and that Genentech just wants to make certain that consumers receive the most accurate information. This is a common tactic for companies caught in misrepresentations: to imply that audience misinterpretation or misunderstanding is a contributing factor to (or the actual cause of) the problem.

"Spinning" a Bad Public-Relations Situation

When Deborah Shank, then a Walmart employee, was severely injured and permanently impaired in a traffic accident in 2000, Walmart paid a total of $470,000 for her ongoing care under the Walmart health-care plan. However, after Mr. and Mrs. Shank won a lawsuit against the trucking company whose vehicle caused the accident, a clause in the Walmart health plan allowed Walmart to sue the Shanks for the reimbursement of the funds that the retailer had expended on her care. After several court actions, Walmart was entitled to recover $275,000, which was all that remained in the fund for Mrs. Shank's ongoing care.

In April 2008, after thousands of online protests against Walmart's policy and actions, the company decided to drop its claim against the Shanks. Walmart Executive Vice President Pat Curran announced the company's change of heart in the following terms, which were reported by CNN:

> Occasionally, others help us step back and look at a situation in a different way. This is one of those times. We have all been moved by Ms. Shank's extraordinary situation. As you know, our current [medical] plan doesn't give us much flexibility, so we began reviewing the guidelines for the trust that pays medical costs for our associates and their family members. . . . We have decided to modify our plan to allow us more discretion for individual cases and are in the final stages of working out the details.
>
> Meanwhile, we wanted you to know that Wal-Mart will not seek any reimbursement for the money already spent on Ms. Shank's care, and we will work with you to ensure the remaining amount in the trust can be used for her ongoing care. We are sorry for any additional stress this uncertainty has placed on you and your family.

As so often happens, the Walmart "spin" didn't enhance the company's reputation. Many observers believed that Walmart should have provided more generous support to the Shanks in the first place. The lesson here is that a company is generally best served when it steps up, apologizes, and takes responsibility for past errors.

E-mailing Your Way to a Legal Loss

In 2007, during an ongoing dispute between staff and management of the *Santa Barbara News-Press*, e-mails written by now-former employees presented a serious setback in efforts to unionize the paper.

Tensions between *News-Press* staff and management began when Wendy McCaw bought the newspaper in 2000. After she took the helm, many editors and other staff of the paper quit or were fired, leading to multiple actions by the National Labor Relations Board (NLRB).

On August 24, 2007, at the height of tensions between the newspaper's employees and management, a group of *News-Press* staffers attempted to deliver a letter of demands to McCaw's office. She was not available, so the letter was delivered to the head of human resources. After the event, now-former employee Tom Schultz sent the following e-mail to a number of recipients, as reported on www.newspress.com:

> Peeps, we rocked the house, crossed their wires and got 'em unglued. Way to go. Anybody feel free to grab me for the full run down on the letter delivery.

Schultz wrote somewhat later:

> Hearing loss . . . must be due to that sonic boom we created during our blitzkrieg through the newsroom on our way to Wendy's office. Dammit, I must have banged on that saucepan too close to my head right before jack-hammering our demands into the floor at HR.

It appears that Schultz was just celebrating and being hyperbolic. However, his e-mails complicated the efforts of the NLRB to prosecute the case against McCaw. The e-mails also prompted McCaw to seek de-certification of the vote by former and current employees to unionize through the Teamsters.

In court, Schultz's e-mails were presented as evidence of workplace disruption by the Teamsters Union and of harassment and intimidation against the owner and the staff still loyal to her.

E-mailing Your Way to Disgrace—and Bankruptcy

With knowledge that his peanut products were tainted with mold toxins, Stewart Parnell, CEO of the Peanut Corporation of America, continued to ship his products to companies in the United States and abroad. This action led to an extensive salmonella outbreak in 2008 and 2009, resulting in nine deaths in the United States and prompting the largest food recall in U.S. history.

Especially interesting to a congressional panel investigating the problem were e-mails from Parnell ordering the tainted shipments. Faced with testing that showed the contamination, Parnell wrote, "We have never found any salmonella at all. No salmonella has been found anywhere else in our products or in our plants." (Parnell's communications were described in the *Huffington Post*, among other sources.) In other e-mails he complained that tests and delays were "costing us huge $$$$$." In an e-mail he wrote to colleagues regarding peanuts that had tested positive for salmonella, he urged, "Turn them loose." Parnell invoked the Fifth Amendment when questioned by the congressional panel.

The Peanut Corporation of America filed for bankruptcy liquidation in February 2009. Parnell and his brother were convicted of serious crimes in September 2014. In September 2015, Stewart Parnell was sentenced in federal court to twenty-eight years in prison; his brother received a twenty-year sentence; and the company's quality assurance manager was sentenced to five years in jail.

Confusing the Issue with Too Many Details

In 2010, many users of Kodak's Easyshare photo-printing service were sent the following merge-list e-mail to explain when their photos would arrive in the U.S. mail. Speedy delivery? Maybe, but don't count on it!

Dear Jon,

Your order from KODAK EASYSHARE Gallery has been completed and is being shipped to the address below. If this message was sent in the evening, over a weekend, or on a holiday, your order will ship on the next business day.

If you'd like to resend this order to friends and family, just visit your order History Page.

3–10 day delivery orders experience a range of delivery times. Most orders are delivered very quickly (within two to five days), but 3–10 day delivery time depends on distance and internal 3–10 day delivery factors, and therefore cannot be guaranteed. U.S. West Coast deliveries often arrive the next day and many U.S. East Coast orders arrive in as few as three days. However, 3–10 day delivery can take as long as 14 days, regardless of destination. If your order has not arrived 14 days after receiving your shipping confirmation e-mail, please contact us at service@kodakgallery.com.

Some orders may be delivered in different shipments depending on the items in the order. To expedite large print orders, delivery will be split into multiple shipments. Multiple shipments will not result in additional shipping charges.

The Kodak communication suggests how too many layers of details can obscure the central message. (Kodak filed for bankruptcy in January 2012 and emerged in September 2013 as a leaner corporation focused on digital imaging. Shutterfly took over Kodak's online photo-printing services.)

Sending an E-mail to the Wrong Recipient— and Jeopardizing a $1 Billion Settlement

In early 2008, the *New York Times* broke the story that an attorney acting on behalf of Eli Lilly & Company mistakenly sent an e-mail to the wrong "Berenson" in her e-mail address book. She wanted to communicate very sensitive negotiations over a possible $1 billion settlement with Eli Lilly Company to her co-counsel Bradford Berenson; instead, she sent her e-mail, shown below, to *Times* reporter Alex Berenson:

> Tom and I were racing to other meetings when we left the EDPA and I am just back, looking for Tom so we can have a call. We'll call you as soon as I have him. Preview: They're in the stratosphere on number and Meehan wants a deal.

Evidently haste makes . . . errors. As John Hutchins explained in 2015 on informationcounts.com, "To the uninformed, the e-mail may be hard to decipher, but Alex Berenson knew exactly what it meant." "EDPA" is the Eastern District of Pennsylvania. Meehan is Patrick—then the U.S. attorney in that district (now a member of Congress). "Meehan wants a deal" and "they're in the stratosphere" means that Eli Lilly and the U.S. attorney were discussing some kind of settlement deal. Reporter Alex Berenson was aware of the talks, but he thought they were on hold. With this new piece of information, he did a little digging, which resulted in a front-page *New York Times* story.

Mass E-mailing Your Way to an Embarrassing Mistake

In March 2009, the Admissions Office at the University of California, San Diego, mistakenly sent a congratulatory "acceptance" e-mail to all 47,000 applicants to UCSD, rather than to the 18,000 students who were actually admitted for the 2009–2010 academic year. According to a *Los Angeles Times* blog, the admissions director and her staff spent many days trying to undo the damage—issuing apologies and responding to angry e-mails and phone calls from parents and students.

In the spring of 2012, UCLA made a similar blunder by sending mistaken admission letters to nearly 900 high school students, as reported by the *Los Angeles Times*.

Mass E-mailing Your Way to a Financial Mess

In December 2011, the venerable *New York Times* sent a 50 percent discount offer to 8.6 million subscribers rather than to the few hundred lapsed subscribers in the newspaper's intended audience. At first, the *Times* honored the offer to all who called in but quickly realized the enormous financial impact this mistaken generosity would have on the company. The *Times* compounded the problem, however, by disclaiming the error to millions of subscribers in a tweet: "If you received an e-mail today about canceling your NYT subscription, ignore it. It's not from us."

But, as the *San Francisco Chronicle* reported on December 29, 2011:

> [T]he *Times* did send the original e-mail, *Times* spokeswoman Eileen Murphy said.
>
> "This e-mail should have been sent to a very small number of subscribers, but instead was sent to a vast distribution list made up of people who had previously provided their e-mail address to the *New York Times*. We regret the error," Murphy said in an e-mail.
>
> "The initial tweet was in error and we regret the mistake," she added.

By the time of this apology, however, the newspaper had received numerous customer complaints and incurred (unspecified) costs because of the mistake.

Making Confidential Business Information Public (via E-mail)

In 2008, the upper-level managers at Carat Media Agency were preparing to lay off some number of employees. They mistakenly sent their confidential planning-process notes to all employees, notes that included all the euphemisms of the process and talking points. As reported by www .adage.com, here is "an unedited passage from the 'message to impacted employees'":

> I unfortunately have some difficult news which affects you and your position with the company. Based on the continued reduction in our client's spend and a restructuring of the core functions (insert group here), we had to evaluate a number of factors and took a hard look at our future and current business need (capacity), performance, and the evolving skill sets needed for our clients and their businesses. As a result, we no longer have a role for you. This was a very difficult decision which is affecting a number of people across Carat. Your last day with the company wil be _____.
>
> I know this is difficult news to handle. I want you to know that we have prepared some information that I would like to review with you now. This is important information concerning your severance, medical benefits, and outplacement assistance. This is the package we have arranged for anyone affected by a reduction in staff such as this.
>
> Please know that we value your contribution to the company and want to help you as your transition into the next stage of your professional career. Let's review your package and make sure you understand what we have provided. We also have outplacement services to offer you as a part of your transition if you are interested in taking advantage of that service.

continued

If you would like to go home today and come back tomorrow to clean out your desk or office, you are free to do so. We would like you to meet with your manager following our meeting to transition your work. We will be communicating to your team today. Your manager will be contacting clients. We ask that you do not contact your client to discuss this situation. As this is affecting a number of people, we will be communicating to the office later today what has occurred.

Please review the materials that we have provided for you. You have one week (or 45 days, depending on situation) to review your severance agreement, sign it, and return it to me. In the meantime, don't hesitate to call me if you have any questions.

You can imagine the reaction from all the employees who received this seeming notice of termination. The misdirected draft also revealed some embarrassing aspects of the managers' planning process.

Making Personal Information Public (via E-mail)

In the summer of 2009, McAfee Inc., the computer-security-software manufacturer, hosted a conference for 800 participants. In a follow-up thank-you e-mail to the conferees, McAfee mistakenly attached a spreadsheet containing personal and professional contact information on the 1,400 people who had attended or expressed interest in the conference. The press delighted in noting the irony of a security-software firm's making a blunder of this magnitude. (One report of the mishap appears on www.techworld.com.au.)

Writing in Code—with Potentially Detrimental Effects

In October 2009, then-Governor Arnold Schwarzenegger sent a message to the California State Assembly vetoing a bill the Assembly had passed. Close readers noted that there seemed to be a hidden F-U message in the memo when one reads the first letter of each line of the two main paragraphs. As reported by the *San Francisco Chronicle*, the governor's aides denied any such intention:

Schwarzenegger's press secretary, Aaron McLear, insisted Tuesday it was simply a "weird coincidence." He sent us veto messages the governor sent out in the past with linguistic lineups such as "soap" and "poet," which he said were also unintended.

"Something like this was bound to happen," McLear said.

You be the judge. Here is the former governor's veto memo:

To the Members of the California State Assembly:

I am returning Assembly Bill 1176 without my signature.

For some time now I have lamented the fact that major issues are overlooked while many unnecessary bills come to me for consideration. Water reform, prison reform, and health care are major issues my Administration has brought to the table, but the Legislature just kicks the can down the alley.

Yet another legislative year has come and gone without the major reforms Californians overwhelmingly deserve. In light of this, and after careful consideration, I believe it is unnecessary to sign this measure at this time.

Sincerely,

Arnold Schwarzenegger

Even if the former governor's veiled message was intentional and not a gaffe, consider the implications of this communication style (along with other Schwarzenegger patterns of behavior) on any future political aspirations he might have.

Posting Your Way to Disgrace

In April 2010, New Jersey teachers protesting against Governor Chris Christie's budget cuts to education lost some credibility when they posted angry, sometimes obscene, and ungrammatical observations on Facebook. Some messages wished that Christie were dead or compared him to Cambodian dictator Pol Pot. As reported on www.nj.com/news, here are some of the teachers' Facebook comments:

"Never trust a fat f---."

"How do you spell A--hole? C-H-R-I-S C-H-R-I-S-T-I-E."

"Remember Pol Pot, dictator of Cambodia? He reigned in terror, his target was teachers and intellectuals. They were either killed or put into forced labor. . . . King Kris Kristy is headed in this direction."

Botching Communications from the Top: Hewlett-Packard

This tale is a good reminder that individual business documents are only part of the larger business writing picture. Major business decisions often require a carefully orchestrated communication strategy involving key players and several interlinked memoranda and announcements.

The Hewlett-Packard (HP) decision in September 2011 to fire CEO Leo Apotheker had numerous causes, but the precipitating cause was Apotheker's communication style. In announcing momentous decisions to restructure the company, including the sudden announcement that HP was likely to sell its personal computer business and discontinue the manufacture of smartphones and tablet computers, Apotheker caught many HP executives off guard and riled HP's corporate customers. In a *New York Times* article, one professor of business characterized the communications as "botched in a big way." The news releases, he said, "came out in dribs and drabs in a very confusing set of announcements."

Botching Communications from the Top: Netflix

Even the media darling Netflix can stumble in communicating its pricing and marketing decisions. For this and other reasons, the company's stock fell 48 percent from July 2011 to September 2011.

In September 2011, Netflix CEO Reed Hastings acknowledged on his blog, "I messed up" in failing to explain persuasively the decision to split (and re-price) its DVD-by-mail and its streaming-video services.

Hastings wrote,

> Most companies that are great at something—like AOL dialup or Borders bookstores—do not become great at new things people want (streaming for us) because they are afraid to hurt their initial business. Eventually these companies realize their error of not focusing enough on the new thing, and then the company fights desperately and hopelessly to recover. Companies rarely die from moving too fast, and they frequently die from moving too slowly. When Netflix is evolving rapidly, however, I need to be extra-communicative. This is the key thing I got wrong. In hindsight, I slid into arrogance based upon past success. . . . I should have personally given a full justification to our members of why we are separating DVD and streaming, and charging for both. It wouldn't have changed the price increase, but it would have been the right thing to do.

Netflix did try to make amends to customers, however, through future communications. On October 11, 2011, the company sent the following e-mail to its many customers. Note that the company even took time to refine its merge list to include only first names (Dear Jon) rather than following the more common corporate style of using full names (Dear Jon Ramsey). Consider the friendly and reassuring tone used by "The Netflix Team":

> Dear Jon,
>
> It is clear that for many of our members two websites would make things more difficult, so we are going to keep Netflix as one place to go for streaming and DVDs.
>
> This means no change: one website, one account, one password . . . in other words, no Qwikster.
>
> While the July price change was necessary, we are now done with price changes.
>
> We're constantly improving our streaming selection. We've recently added hundreds of movies from Paramount, Sony, Universal, Fox, Warner Bros., Lionsgate, MGM, and Miramax. Plus, in the last couple of weeks alone, we've added over 3,500 TV episodes from ABC, NBC, FOX, CBS, USA, E!, Nickelodeon, Disney Channel, ABC Family, Discovery Channel, TLC, SyFy, A&E, History, and PBS.
>
> We value you as a member, and we are committed to making Netflix the best place to get your movies & TV shows.
>
> Respectfully,
>
> The Netflix Team

Making (Expensive) Punctuation Mistakes

In 2006, a contract dispute arose in Canada between Rogers Communications and Bell Aliant over just when their contract could be terminated "by either party." The confusion hinged on the inclusion of one (extra?) comma after the word "terms" in this part of the agreement between the two companies:

> [This agreement] shall be effective from the date it is made and shall continue in force for a period of five (5) years from the date it is made, and thereafter for successive five (5) year terms, unless and until terminated by one year prior notice in writing by either party.

As reported on NPR, Bell Aliant wanted to terminate the contract short of five years just by giving "one year prior notice." Despite what most close readers would see as the obvious intent of the two parties to enter into a long-term contract, the Canadian Radio-television and Telecommunications Commission decided in favor of Bell Aliant and allowed its early release from the contract.

Making Serious Social-Media Errors

Clearly such social media as Twitter, Facebook, Pinterest, and LinkedIn are playing a prominent role in business communications—thus far primarily for public-relations and marketing purposes. While marketing strategies are not a focus of *Business Writing Scenarios*, it's useful for you to consider how easily an ill-conceived tweet or an angry post to Facebook can damage your career, the company for which you work, or both. Some of the cyberspace embarrassments and disasters brought to our attention daily turn out to be fake events manufactured by hackers and others, but many others are actual errors of individual or corporate judgment that are disseminated to huge communities of people around the world. Gathered here are a few snapshots of what can go wrong in social-media communications.

Complaining about a Customer through Reddit

In January 2013, an Applebee's waitress was insulted when a large group left her with no tip and when a member of the group wrote this on the receipt:

I give God 10%, why do you get 18

The customer was unhappy about the automatic 18 percent gratuity added for her large group. The displeased waitress posted a picture of the receipt online through Reddit and commented:

> Parties up to eight . . . may tip whatever they'd like, but larger parties receive an automatic gratuity. It's in the computer, it's not something I do. They had no problem with my service, and told me I was great. They just didn't want to pay when the time came.

The customer's name was visible on the posted receipt, and the post went viral on the Internet. The customer, a pastor at Truth in the World Deliverance Ministries Church, was embarrassed by the notoriety and complained to Applebee's management. The waitress was fired for leaking a professional matter at Applebee's to a huge social-media audience.

Complaining about a New Job on Twitter

In 2009, a young woman newly hired for an internship by Cisco in San Jose, California, complained in a Twitter message that she would hate making the long commute and hate her work as well:

> Cisco just offered me a job! Now I have to weigh the utility of a fatty paycheck against the daily commute to San Jose and hating the work.

Of course, Cisco (like many companies) trolls the Internet for information related to its brand and company interests, and quickly the Twitter message was discovered. A Cisco advocate responded:

> Who is the hiring manager? I'm sure they would love to know that you will hate the work. We here at Cisco are versed in the web.

A bit of sleuthing quickly revealed the identity of the new intern, who became the butt of many online jokes, criticisms, and parodies, even engendering a Web site called CiscoFatty.com. The employee's internship offer was quickly rescinded by Cisco. Ironically, the author of the perilous tweet was working on a master's degree in information management and systems. In an interview with Technotica, part of www.nbcnews.com, she reflected:

> I was using Twitter in a way that didn't jibe with how Twitter really works. I was using it more like I was on Facebook. I was posting status updates to people who are my friends, not realizing or caring that everybody in the whole world could see my updates because I wasn't thinking my updates were interesting to anybody outside my group. Yup, I certainly learned the hard way.

Insulting a Key Client (and Others) on Twitter

On his way to an important meeting with the CEO of FedEx, a vice president at Ketchum Inc. foolishly tweeted his aversion to the city where the meeting was being held—Memphis, Tennessee:

> True confession but i'm in one of those towns where I scratch my head and say "I would die if I had to live here!"

Ketchum is a global public-relations firm, and FedEx was (and remains) a major client for Ketchum's services. Internet cruisers very quickly spotted the insult and sent it to executives at both FedEx and Ketchum. A long series of rejoinders and apologies followed. One defender of FedEx and of Memphis wrote:

> If I interpret your post correctly, these are your comments about Memphis a few hours after arriving in the global headquarters city of one of your key and lucrative clients, and the home of arguably one of the most important entrepreneurs in the history of business, FedEx founder Fred Smith.

FedEx executives also provided this important reminder for all of us to consider as we use social media:

> This lapse in judgment also demonstrates the need to apply fundamental communications principles in the evolving social networking environment: Think before you speak; be careful of what you say and how you say it. Mr. Andrews [the Ketchum vice president] made a mistake, and he has apologized. We are moving on.

Mixing Personal Expressions with Company Communications

During the presidential debates of 2012, an employee at KitchenAid who was responsible for social-media advertising mistakenly sent this personal tweet through the company's system: